Mastering Text Analytics

A Hands-on Guide to NLP Using Python

Shailendra Kadre
Shailesh Kadre
Subhendu Dey

Mastering Text Analytics: A Hands-on Guide to NLP Using Python

Shailendra Kadre
RBI Layout, Carinata Block B-311
Brigade Gardenia
Bangalore, Karnataka, India

Shailesh Kadre
Hyderabad, Telangana, India

Subhendu Dey
Bangalore, Karnataka, India

ISBN-13 (pbk): 979-8-8688-1581-2
https://doi.org/10.1007/979-8-8688-1582-9

ISBN-13 (electronic): 979-8-8688-1582-9

Copyright © 2025 by Shailendra Kadre, Shailesh Kadre and Subhendu Dey

This work is subject to copyright. All rights are reserved by the Publisher, whether the whole or part of the material is concerned, specifically the rights of translation, reprinting, reuse of illustrations, recitation, broadcasting, reproduction on microfilms or in any other physical way, and transmission or information storage and retrieval, electronic adaptation, computer software, or by similar or dissimilar methodology now known or hereafter developed.

Trademarked names, logos, and images may appear in this book. Rather than use a trademark symbol with every occurrence of a trademarked name, logo, or image we use the names, logos, and images only in an editorial fashion and to the benefit of the trademark owner, with no intention of infringement of the trademark.

The use in this publication of trade names, trademarks, service marks, and similar terms, even if they are not identified as such, is not to be taken as an expression of opinion as to whether or not they are subject to proprietary rights.

While the advice and information in this book are believed to be true and accurate at the date of publication, neither the authors nor the editors nor the publisher can accept any legal responsibility for any errors or omissions that may be made. The publisher makes no warranty, express or implied, with respect to the material contained herein.

Managing Director, Apress Media LLC: Welmoed Spahr
Acquisitions Editor: Celestin Suresh John
Coordinating Editor: Gryffin Winkler

Cover image designed by Freepik (www.freepik.com)

Distributed to the book trade worldwide by Springer Science+Business Media New York, 1 New York Plaza, New York, NY 10004. Phone 1-800-SPRINGER, fax (201) 348-4505, e-mail orders-ny@springer-sbm.com, or visit www.springeronline.com. Apress Media, LLC is a Delaware LLC and the sole member (owner) is Springer Science + Business Media Finance Inc (SSBM Finance Inc). SSBM Finance Inc is a **Delaware** corporation.

For information on translations, please e-mail booktranslations@springernature.com; for reprint, paperback, or audio rights, please e-mail bookpermissions@springernature.com.

Apress titles may be purchased in bulk for academic, corporate, or promotional use. eBook versions and licenses are also available for most titles. For more information, reference our Print and eBook Bulk Sales web page at http://www.apress.com/bulk-sales.

Any source code or other supplementary material referenced by the author in this book can be found here: https://www.apress.com/gp/services/source-code.

If disposing of this product, please recycle the paper

This book is dedicated to my father, Sharad Madhav Kadre; my mother, Shakuntala Kadre; my wife, Meenakshi Karkare; my daughter, Neha Kadre, and my son, Vivek Kadre

—*Shailendra Kadre*

To my wife, Neena Kadre, and my daughters, Nandini and Samrudhi

—*Shailesh Kadre*

To my late father, Nitish Chandra Dey; my late mother, Dipika Dey; my wife, Chaitali Dey; and my son, Subhashish Dey

—*Shubhendu Dey*

Table of Contents

About the Authors xv

About the Technical Reviewer xix

Acknowledgments xxi

Introduction xxiii

Chapter 1: Natural Language Processing: An Introduction 1

1.1 Why You Should Read This Chapter 1

1.2 Introduction to Natural Language Processing 2

    1.2.1 NLP Techniques 5

    1.2.2 Text Classification 5

    1.2.3 Clustering 7

    1.2.4 Collocation 9

    1.2.5 Concordance (Computing Word Frequencies) 10

    1.2.6 Text Extraction 10

    1.2.7 Stop Word Removal 11

    1.2.8 Word Sense Disambiguation (WSD) 12

1.3 Challenges in NLP 14

1.4 Applications of NLP 16

    1.4.1 Banking and Financial Industry (BFSI) 18

    1.4.2 Healthcare 19

    1.4.3 Legal 19

TABLE OF CONTENTS

1.4.4 Automate Routine Business Processes 20
1.4.5 Improve Search 20
1.4.6 Search Engine Optimization 21
1.4.7 Machine Translation: Translating Languages Automatically Using NLP 22
1.4.8 Text Summarization: Condensing Large Text Information 23
1.5 Recent Trends and Future Directions 24
   1.5.1 Transfer Learning: Advantages of Pre-Training 25
   1.5.2 LLMs: Human-like Interactions 26
   1.5.3 Cross-Lingual and Multiple-Lingual Models: Handling Multiple Languages 27
   1.5.4 Explainable NLP: Reason Model Results 28
   1.5.5 Emergent Areas in NLP 29
1.6 Short Business Case 1: Application of NLP in Medical Text Analysis 30
   1.6.1 Streamlining Clinical Research 30
   1.6.2 Improving Administrative Efficiency 31
   1.6.3 Enhancing Data Accessibility 31
   1.6.4 Facilitating Predictive Analytics 32
   1.6.5 Business Case Conclusion 32
1.7 Short Business Case 2: NLP in Customer Service, Enhancing Customer Interaction 33
   1.7.1 The Challenge 34
   1.7.2 Implementation of NLP 34
   1.7.3 Results and Benefits 36
   1.7.4 Conclusion 36
1.8 Example with Python 37
   1.8.1 NLP Tutorial 37
   1.8.2 YouTube Comments Spam Detection Using Python 37

1.9 Recap of Key Concepts 46

1.10 Conclusion 47

1.11 Exercises 47

1.12 References 49

Chapter 2: Collecting and Extracting the Data for NLP Projects 53

2.1 Why You Should Learn NLP 53

2.2 Where to Find the Data (Text Corpora) for NLP Projects 54

    2.2.1 Accessing Some Popular NLTK Text Corpora 57

    2.2.2 Accessing Gutenberg Corpus 58

    2.2.3 Accessing Reuters Corpus 58

    2.2.4 Accessing Brown Corpus 59

    2.2.5 Accessing Gutenberg Corpus 60

    2.2.6 Accessing Web and Chat Text 60

    2.2.7 Accessing NLTK Lexical Resources 61

2.3 Extracting Data from Word Files 63

    2.3.1 Extracting Data from MS Word Files 64

2.4 Extracting Data from HTML 67

    2.4.1 Extracting Data from HTML Documents 68

2.5 Extracting Data from JSON 72

    2.5.1 Extracting Data from JSON Files 72

2.6 Extracting Data from PDFs 75

    2.6.1 Extracting Data from JSON Files 76

2.7 Web Scraping 78

    2.7.1 Web Scraping in NLP Tasks 79

2.8 Recap of Key Concepts 80

2.9 Practice Exercises 81

2.10 References 81

TABLE OF CONTENTS

Chapter 3: NLP Data Preprocessing Tasks Involving Strings and Python Regular Expressions 83

3.1 Why You Should Read This Chapter..83
3.2 Python for Language Processing 84
3.2.1 A Text Analytics Project Life Cycle (Generic NLP Pipeline)...................85
3.2.2 String Handling in Python..88
3.2.3 Introducing Regular Expressions for Text Processing.........................91
3.3 Real-life (NLP) Applications of RegEx...94
3.3.1 Validate Formatting of Phone Numbers 95
3.3.2 Search and Replace...96
3.3.3 Date Formatting..97
3.3.4 Word Counting 97
3.3.5 Log File Analysis 99
3.3.6 String Cleaning 100
3.3.7 Tokenization..101
3.3.8 Text Normalization 103
3.4 Chapter Recap 105
3.5 Exercises...105
3.6 References...106

Chapter 4: NLP Data Preprocessing Tasks with NLTK 107

4.1 Why You Should Read This Chapter..107
4.2 NLP Data Preprocessing Tasks...108
4.2.1 Tokenization Using NLTK..108
4.2.2 Stop Word Removal 110
4.2.3 Stemming Using NLTK 113
4.2.4 Lemmatization..116
4.2.5 Sentence Segmentation 119

4.2.6 Word Frequency Distribution 121

4.2.7 Antonym and Synonym Detection Using NLTK 124

4.2.8 Word Similarity Calculation 125

4.2.9 Word Sense Disambiguation Using NLTK 127

4.2.10 Keyword Extraction with NLTK 129

4.3 NLP Tasks Beyond Preprocessing 132

4.3.1 Part-of-Speech Tagging 132

4.3.2 Named Entity Recognition with NLTK 135

4.4 Some Useful Functionalities Outside NLTK 138

4.4.1 Word Cloud Generation 139

4.4.2 Language Translation 142

4.4.3 Text Summarization 144

4.5 Chapter Recap 147

4.6 Exercises 148

Chapter 5: Lexical Analysis **149**

5.1 Why You Should Read This Chapter 149

5.2 Morphological Analysis Revisited 150

5.3 Tokenization Revisited 151

5.3.1 Code Demos 152

5.4 Stemming Revisited 166

5.4.1 Code Demo for Multiple Stemmers 167

5.5 Lemmatization Revisited 171

5.5.1 Code Demo for Multiple Lemmatizers 172

5.6 A Complete Lexical Analysis with Data Preprocessing 176

5.6.1 YouTube Comments Spam Detection 176

5.6.2 Convert the Text to Lowercase 179

5.6.3 Remove All Unwanted Punctuation 180

TABLE OF CONTENTS

    5.6.4 Remove Stop Words 181
    5.6.5 Remove Frequent Words 183
    5.6.6 Remove Rare Words 184
    5.6.7 Stemming 186
    5.6.8 Lemmatization 189
    5.6.9 Removal of Emojis 192
    5.6.10 Removal of Emoticons 194
    5.6.11 Removal of URLs 199
    5.6.12 Removal of HTML tags 200
    5.6.13 Chat Words Conversion 202
    5.6.14 Spelling Corrections 204
    5.6.15 Apply Machine Learning Model 206
5.7 Chapter Recap 208
5.8 References 209

Chapter 6: Syntactic and Semantic Techniques in NLP 211
6.1 Why You Should Read This Chapter 211
6.2 An Overview of Structural Techniques 213
    6.2.1 Part-of-Speech Tagging 214
    6.2.2 POS Tagging Tutorials 217
    6.2.3 Parsing Techniques 220
    6.2.4 Hands-on Example Covering for Parsing 225
    6.2.5 Challenges in Syntactic Analysis 228
6.3 Introduction to Semantics 229
    6.3.1 Chunking in NLP 230
    6.3.2 Named Entity Recognition 233
    6.3.3 WSD Revisited 234
    6.3.4 Term-Document Matrix (Co-Occurrence Matrix) 236

TABLE OF CONTENTS

    6.3.5 Term Frequency–Inverse Document Frequency 238

    6.3.6 Coding Tutorials for This Section 242

6.4 Lexical Semantics 247

    6.4.1 Synonyms 248

    6.4.2 Antonyms 249

    6.4.3 Homophones 251

    6.4.4 Homographs 252

    6.4.5 Polysemy 253

    6.4.6 Hyponyms 254

    6.4.7 Building Lexical Semantic Models with Code Examples 255

    6.4.8 Integration into Text Processing Pipelines 267

6.5 Chapter Recap 272

6.8 References 274

Chapter 7: Advanced Pragmatic Techniques and Specialized Topics in NLP 275

7.1 Why You Should Read This Chapter 275

7.2 Overview of Pragmatic Analysis Techniques 277

7.3 Discourse Integration in NLP 277

    7.3.1 Techniques for Discourse Integration 278

    7.3.2 Demonstrating Techniques for Discourse Integration 280

7.4 Distributional Semantics and Word Embeddings 307

    7.4.1 Latent Semantic Analysis 310

    7.4.2 Popular Word Embeddings 313

    7.4.3 Case Studies and Applications 331

7.5 Summary 335

7.6 Reference 335

TABLE OF CONTENTS

Chapter 8: Transformers, Generative AI, and LangChain 337

8.1 Introduction to Transformers in NLP 337
    8.1.1 Evolution of NLP: From Traditional Models to Transformers 340
    8.1.2 Overview of BERT by Google 342
    8.1.3 Code to Demonstrate the Use of DistilBERT Using PyTorch................343
    8.1.4 Multimodal NLP: Combining Text with Images, Audio, and More........350
    8.1.5 Demonstrating the Use of OpenAI CLIP for Multimodal Understanding 351
    8.1.6 Text-to-Image Search Using OpenAI's CLIP Model 353

8.2 Overview of GPT 356
    8.2.1 Evolution of Text Generation Models: From RNNs to GPT 356
    8.2.2 A Short Note on the Working of GenAI (LLMs) Models 358
    8.2.3 Applications and Future Trends in Generative AI 359

8.3 LangChain and OpenAI's (GenAI) APIs 361
    8.3.1 Setting Up the Environment..363

8.4 Model I/O: Easily Interface with Language Models 364
    8.4.1 Using LLMs with LangChain 370
    8.4.2 Using Chat Models with LLMs 372
    8.4.3 Working with Prompt Templates and Few-Shot Templates 385
    8.4.4 Parsing and Serialization..390
    8.4.5 Customizing Model Inputs and Handling Outputs..............................400
    8.4.6 Switching Between Different LLMs 405

8.5 Summary..408

Chapter 9: Advancing with LangChain and OpenAI 411

9.1 Data Connection with Application-Specific Data Sources........................412
    9.1.1 Document Loading, Transformation, and Integration........................412
    9.1.2 Chains: Construct Sequences of Calls for Specific Tasks 419
    9.1.3 Text Embeddings and Vector Databases..425

TABLE OF CONTENTS

9.2 AI Agents 430

    LangChain-Based Stock Analysis Agent 430

    The Role of LLM and Agents in This Code 436

9.3 Summary 437

Chapter 10: Case Study on Symantec Analysis 439

10.1 Introduction 439

10.2 Methods of Semantic Analysis 441

    10.2.1 Latent Semantic Analysis 442

10.3 Role of Semantic Analysis in Enhancing Customer Experience:
A Case Study Fujitsu's Kozuchi AI Agent 453

    10.3.1 About Fujitsu 453

    10.3.2 AI Initiatives of Fujitsu 455

    10.3.3 Fujitsu Kozuchi for Semantic Analysis 457

    10.3.4 Business Applications of Fujitsu Kozuchi and Its Superiority over Other Similar Tools 459

10.4 Future Trends in Semantic Analysis 460

Conclusion 462

10.5 References 462

Index 465

About the Authors

Shailendra Kadre is a seasoned professional in machine learning, deep learning, natural language processing, generative AI, and digital transformation, with 30 years of industry experience at top-notch IT products and services companies across the globe. He has held leadership positions in AI and Product Analytics at HP Inc. and Satyam.

Shailendra is currently with Christ University, Bangalore, in an academic position. Shailendra holds a master's degree in design engineering from the prestigious Indian Institute of Technology (IIT), Delhi, and a master of science in artificial intelligence (AI) and machine learning (ML) from Liverpool John Moore's University, UK. He is also a certified Project Management Institute (PMI) Project Management Professional (PMP).

Shailendra is the author of books like *Going Corporate* (Apress, 2011), *Practical Business Analytics Using SAS* (Apress, 2015), and *Machine Learning and Deep Learning using Python and TensorFlow* (McGraw-Hill, 2021). All of his books have received positive reviews on Amazon. A photography enthusiast and an author, Shailendra has successfully exhibited his landscape photographs. He resides in Bangalore, India, with his family.

ABOUT THE AUTHORS

Dr. Shailesh S. Kadre is a seasoned professional with over 20 years of experience in engineering analysis, project management, and product design across aerospace, railways, automotive, heavy engineering, and consumer product domains. Holding a PhD in optimization algorithms, a master's of science in AI/ML, and a master's of technology from IIT Kharagpur, he currently serves as senior divisional manager at Cyient Ltd., Hyderabad, specializing in structural analysis for aero-engine components and railway systems.

Dr. Kadre's contributions include optimizing submodeling techniques for aero-engine automotive structures. He has also developed surrogate model-assisted optimization techniques to enhance the efficiency of finite element optimization processes. Additionally, he implemented optimization processes for railway structures to achieve weight reductions and developed shock and vibration analysis procedures compliant with industry standards, eliminating the need for physical testing in specific scenarios.

Dr. Subhendu Dey serves as the Pro Vice Chancellor of Sister Nivedita University in Kolkata, India. A distinguished academic leader, he brings extensive expertise as an educator, researcher, administrator, and institution builder. His academic and research focus spans marketing analytics and subsistence marketplaces, with an emphasis on small and marginal agri-fresh producers, as well as sustainability management. His scholarly contributions include over 30 research papers and more than a dozen

ABOUT THE AUTHORS

case studies on Indian businesses, published in leading national and international journals. Additionally, he has provided consultancy services to organizations in India, the Kingdom of Saudi Arabia, Russia, and Ukraine. Beyond academia, Subhendu's interests lie in acting, writing fiction novels, and philanthropic works. He resides in Kolkata, India, with his family.

About the Technical Reviewer

Milind Kolhatkar is a seasoned expert in financial markets and technology with over 30 years of experience. He specializes in treasury, derivatives, and capital markets, bringing strong capabilities in software product management, marketing, lifecycle management, presales, and customer engagement.

Milind currently serves as a product manager at a leading financial products firm. He holds a bachelor's of technology from IIT Bombay and an MBA from IIM Ahmedabad. He is also an avid follower of developments in artificial intelligence, particularly its impact on software development, requirements management, and customer support.

Beyond his professional interests, Milind enjoys long-distance running, exploring diverse cultures, and studying world history.

Acknowledgments

This book is acknowledged to our numerous friends, well-wishers, readers, and professionals, whose help and feedback made possible the creation of this book. We acknowledge the help extended by Apress editors Celestin Suresh John and Gryffin Winkler, without whom this book would not be possible at all. We also acknowledge the help of Apress copy editors and several other Apress operations staff, whose timely assistance was instrumental in bringing this book to its current shape.

Introduction

This book is for all professionals, academics, and students who want to learn and master hands-on skills in the domains of natural language processing (NLP) and generative AI (GenAI).

This book assumes its readers have intermediate skills in machine learning, deep learning, and, most importantly, Python. The book begins with the basics and gradually progresses to advanced concepts in transformer models and generative AI.

The authors expect you to read this book with a Python editor on a decent computing device, preferably with 8 GB RAM and I5 (or above) or equivalent CPU. It's a hands-on book, which will benefit you only if you program the code snippets given in the book as you go along.

All the code and datasets used in this book are available at `https://github.com/Apress/Mastering-Text-Analytics`.

CHAPTER 1

Natural Language Processing: An Introduction

1.1 Why You Should Read This Chapter

This chapter serves as an introduction to the rest of the book, which explores language processing with the aid of machines, such as computers and robots. Why is natural language processing (NLP) important?

NLP can analyze language for its meaning. Here, by *language*, we mean the lingos like English, German, French, and Hindi, which humans use in their day-to-day transactions. NLP, with the aid of machines (primarily computer systems), has long fulfilled useful roles. Such systems can correct grammar, convert speech to text, and even translate between languages. Google Translator is one of the best examples of such translators. NLP can help machines communicate with humans in their language. With the help of this technology, machines can automatically read text, hear and interpret speech, and perform public sentiment analysis (on any topic, such as new product introductions and politics) on social media platforms like Twitter (now X).

CHAPTER 1   NATURAL LANGUAGE PROCESSING: AN INTRODUCTION

1.2 Introduction to Natural Language Processing

It's time to formally introduce NLP. Amazon defines it as "a machine learning technology that gives computers the ability to interpret, manipulate, and comprehend human language." Similarly, IBM defines it as "a subfield of computer science and artificial intelligence (AI) that uses machine learning to enable computers to understand and communicate with human language."

Research in the field of NLP has made possible the new era of generative AI (GenAI) platforms like ChatGPT and many others. These types of software are called large language models (LLMs). GenAI) software can generate text, images, and even videos. NLP is already a routine part of life for many. This kind of users are not only concentrated in developed countries, but many developing countries across the globe can't imagine their day-to-day working without NLP and GenAI software. Some common applications of NLP are powering search engines like those from Google and Microsoft, prompting chatbots used for customer service, speech-based GPS systems, and even digital assistants like Siri by Apple and Amazon's Alexa.

The comparatively recent advances in the fields of data sciences and machine learning have greatly benefited the art and science of NLP. The field of NLP is largely divided into three areas.

- **Understanding speech**: The translation of spoken languages like English and many others into human-readable text.

- **Understanding natural languages**: A machine's (usually computers or equivalent) capability to understand language.

- **Generation of natural languages:** This points us to ChatGPT-like applications that make natural language generation by computers.

CHAPTER 1   NATURAL LANGUAGE PROCESSING: AN INTRODUCTION

The processing of natural languages used by humans is special for many reasons. First of all, the meaning of a word in a sentence or a paragraph is contextual, meaning the same word can have different meanings in different contexts.

According to Chris Manning, an AI professor at Stanford, NLP is a signaling system where symbols are discrete, symbolic, and categorical in nature. The same meaning can be conveyed in different ways. It can be speech, gesture, or signs. Each of these ways is unique and ambiguous. Other challenges that make it difficult are misspelled words, colloquialisms and slang, too many local languages and dialects (like there are over 3,000 languages in the African continent alone), and so on.

While the science of NLP has its limitations, it still offers widespread benefits to humanity. With rapid developments in new technologies, many of these blocks are getting removed at an unprecedented rate.

Let's consider the amount of data generated globally. Srinivasa-Desikan gives some concrete data: Google handles more than 1 trillion queries per year), Twitter has 1.6 billion queries every day, and WhatsApp handles more than 30 billion posts per day. Let's look at the following figures.

- About 402.74 million terabytes of data are produced worldwide each day

- Approximately 147 zettabytes of data were produced in 2024

- Roughly 181 zettabytes of data will be produced in 2025

Figure 1-1 highlights the amount of data every year since 2020. Needless to say, much of this data consists of both structured and unstructured speech data. Companies generate immense value by analyzing this data using various machine learning techniques. They are valuable business insights that can be used in data-driven business

CHAPTER 1   NATURAL LANGUAGE PROCESSING: AN INTRODUCTION

decision-making. Many machine learning techniques for analyzing data are bundled under the banner of NLP, which is used to automatically extract valuable insights from both structured and unstructured text data. Companies utilize text analysis techniques to swiftly digest online data and documents generated and collected from numerous sources and transform them into actionable insights. Global tech companies like Amazon, Google, and Microsoft have taken a considerable lead over others in this direction.

*Machine learning is a developing branch of computer algorithms that can emulate human intellect by learning from the neighboring environment (both text and numeric data). [6]

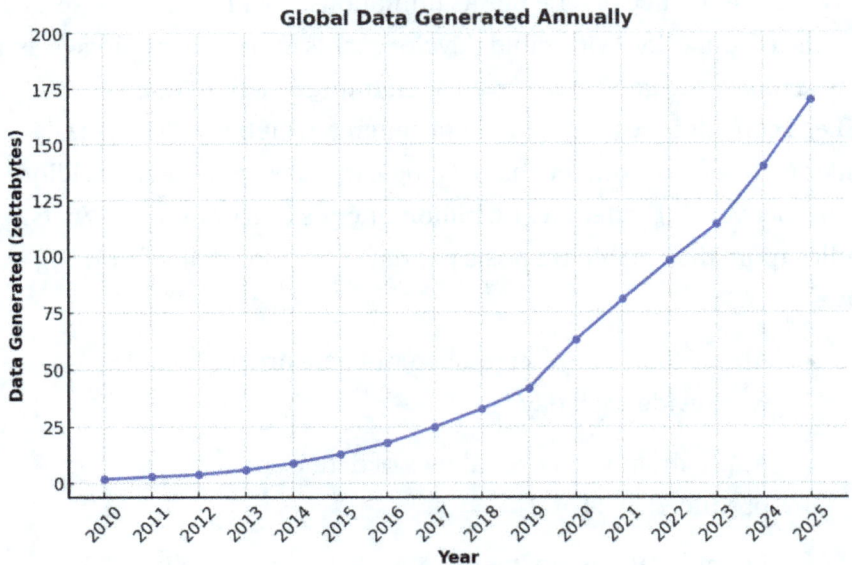

Figure 1-1. *Global data generation 2010 to 2025*

NLP is discussed in the following sections.

1.2.1 NLP Techniques

More formally, NLP is extracting high-quality and actionable information from text data. A vast amount of text data is available on websites, emails, reviews like those on Amazon and X (formerly Twitter), books, and research papers. This data is extracted, cleaned, formatted, and undergoes other data processing steps. This data processing is a laborious process, and it can account for up to 80% of the total time spent on the entire NLP project. Then, finally, statistical algorithms are used to derive patterns within the text data. Both machine learning and deep learning algorithms are utilized depending on the task at hand. The results are finally evaluated and interpreted for use in data-driven decision-making. Typical NLP tasks comprise text categorization and clustering, entity extraction, development of granular catalogs, text summarization, and many others. Sentiment analysis of social media data is one of the popular tasks performed using NLP techniques.

NLP uses many basic advanced techniques. Simpler ones include text classification, computing word frequencies, text extraction and clustering, part-of-speech tagging, sentiment analysis, collocation and concordance, stop word removal, and word sense disambiguation. Next, let's briefly look at each of these simpler techniques. We examine more complex NLP techniques in later chapters.

1.2.2 Text Classification

Text classification is an analytical method used by professionals to categorize or sort text into different predefined groups. For example, a tweet can represent a positive or negative sentiment (positive is represented as 1 and negative as 0). Now, suppose you have one thousand such tweets, all prelabelled labeled as either positive or negative. Now, you possibly train a machine simple machine learning algorithm, logistic regression. Though only one thousand tweets can be slightly insufficient data to properly train any machine learning algorithm, it can still work as a demonstrative example.

CHAPTER 1   NATURAL LANGUAGE PROCESSING: AN INTRODUCTION

Once the logistic regression model is trained, it can be used to classify any new tweet(s) as positive or negative. This type of project comes under the category of text classification. Logistic regression belongs to the linear family of machine learning algorithms, and it is probably the simplest in our kit of such tools. For completeness, there are other non-linear and more complex classification algorithms like random forests, XGBoost, and those from the neural networks family. All these complex algorithms can be deployed in text classification projects, provided the data is properly cleaned and preprocessed to the requirements of such text classification projects. These non-linear and more complex algorithms can be more accurate than logistic regression in some cases. The choice of algorithms depends upon many factors, like the nature of the task, availability of computing resources, skill availability, and available time. Figure 1-2 depicts the working of a simple email classifier.

CHAPTER 1   NATURAL LANGUAGE PROCESSING: AN INTRODUCTION

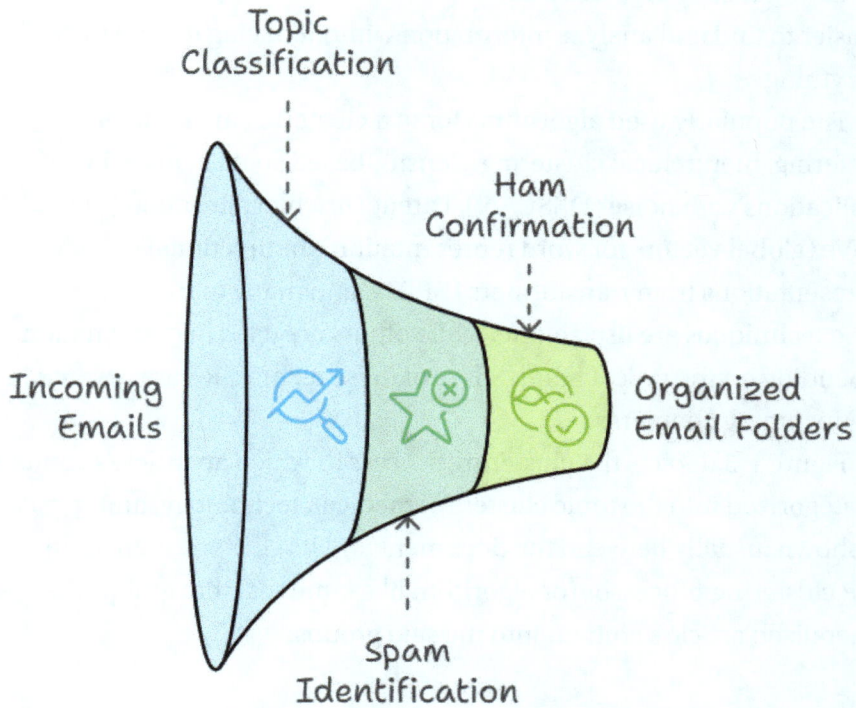

Figure 1-2. Topic classification is used to flag incoming spam emails, which are filtered into a spam folder

1.2.3 Clustering

Let's try to understand clustering with a simple example. Imagine you have ten different text paragraphs about a variety of topics. As of now, you are not given the titles of any of the text paragraphs. While reading, you discover that some are about linear algebra, some about the history of Europe, and some about quantum mechanics. You group the text paragraphs based on these topics. That's a perfect example of text clustering!

CHAPTER 1   NATURAL LANGUAGE PROCESSING: AN INTRODUCTION

Having learned about text clustering, now let's look at an example of how it can help us. Imagine you have managing large amounts of text data whose titles are known. Text clustering using machine learning algorithms can automatically organize such data into meaningful groups. This makes it easier to find and analyze information without having to read through everything.

The popularly used algorithms for text clustering are k-means clustering, hierarchical clustering, density-based spatial clustering of applications with noise (DBSCAN), Latent Dirichlet allocation (LDA), GloVe (global vectors for word representation), bidirectional encoder representations from transformers (BERT), and many more. Some of these techniques are used alone, while others are used in combination, depending on the task at hand. Some of these techniques are covered in the upcoming chapters.

Figure 1-3 depicts the clustering process in which an article's content is categorized into the topic clusters of medical, technology, and sports. As shown, usually between the document and its clusters, there is an NLP clustering processor (or algorithm like k-means) that groups the generalized article's content into the said groups.

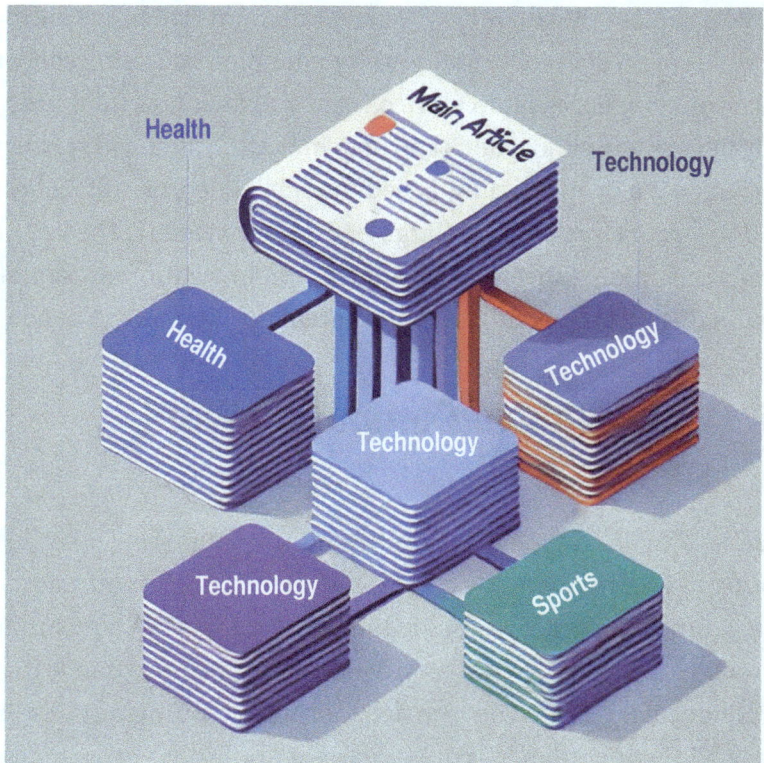

Figure 1-3. *Topic-wise clustering of a general article*

1.2.4 Collocation

Let's again start with a simple example. Suppose you are a cook in a five-star hotel. Over the years of cooking, you noticed that certain ingredients are often used together in recipes. Examples of such ingredients include "peanut butter" and "jelly," or "salt" and "pepper". Such pairs of ingredients are called *collocations* in the cooking domain. When it comes to text or language, collocations are pairs or groups of words that often go together and sound natural to native speakers.

CHAPTER 1   NATURAL LANGUAGE PROCESSING: AN INTRODUCTION

"Make a choice", "Take a break", "Heavy rain", and "Do homework" are a few examples of English collocations. Understanding and using collocations can make your language sound more natural and fluent. As the words naturally fit together, collocations can help in both writing and speaking.

Word2Vec is a deep learning-based model that learns collocation word associations from large corpora (large bodies of text data). The key here is that words that often appear together have similar vector representations. We discuss this and many other techniques in the chapters to come.

1.2.5 Concordance (Computing Word Frequencies)

As usual, let's again start with a simple example. Suppose you are attempting to write a relatively long article. In this article, you want to see how many times a specific word, say rain, has appeared. A concordance helps you do just that. The entire process of finding concordance (or word frequencies) is done with the help of computer programs. So, in the context of NLP, concordance is the process of finding the occurrences of a word or phrase in a text corpus, along with their contexts.

NLP-specific Python libraries like Natural Language Toolkit (NLTK) can be used to create concordances by processing and analyzing text data. We discuss this library in the early part of this book. Some other software tools are also capable of providing visual representations of word frequencies and their contexts.

1.2.6 Text Extraction

In the context of NLP-specific computer programs, text extraction simply means pulling out specific information of interest from a large amount of text corpus. Let's try to understand with the help of a simple example.

Imagine you are reading a book and you only need the paragraphs that discuss your hobby, such as tennis. Text extraction is like using a highlighter to mark just those parts. Under text extraction, the main topics that are covered in this book are as follows.

- **Named entity recognition (NER)**: Extracting various entities like names, dates, locations, and other specific items from a text corpus.

- **Keyword extraction**: Extracting important keywords or phrases that summarize the main theme of the document or text corpus.

- **Part-of-speech tagging**: Labeling words in a text corpus, including various nouns, verbs, and adjectives. It helps in understanding the structure and extracting relevant information.

- **Regular expressions (RegEx)**: Using patterns to extract explicit text strings, such as finding addresses or phone numbers from the body of emails.

- **Text summarization**: Producing a summary of a longer text document and highlighting the key points.

1.2.7 Stop Word Removal

As usual, let's again start with a simple example. Consider the sentence, "The cat is on a mat." If all the stop words are removed, we are left with only "cat" and "mat," which are the keywords. Stop word removal is a process used in NLP that filters out commonly occurring words. The type of words that are removed can be "and", "the", "is", "in", and so forth. These kinds of words don't add much meaning to sentences. We call words like "and," "the," "is," and "in" *stop words* in the context of NLP. Removing these words helps computers focus on the more important words in a text corpus.

The stop word removal process begins by tokenizing the text. Tokenizing is breaking a text document (a document in the context of NLP is a complete piece of text) into individual words. The next step is to compare each word against a predefined list of stop words. If any word matches one on the list, it is to be removed from the text. This step helps reduce the dataset size (of individual words) and keeps the focus of analysis on the more significant words.

Popular libraries for stop word removal in NLP contain NLTK (Natural Language Toolkit), spaCy, Gensim, scikit-learn, and TextBlob. These libraries offer comprehensive functions for general text processing, including stop word removal. We introduce most of these libraries in the early part of this book. These libraries are utilized throughout this book when we discuss our case studies with code.

1.2.8 Word Sense Disambiguation (WSD)

Consider the simple sentence: "John noticed a bat flying in the sky." Here, the word "bat" could have two distinct meanings. The first one is the nocturnal animal, and the second is equipment used in baseball, tennis, or cricket. WSD program analyzes the words surrounding "bat" and the syntactic structure of the entire sentence. It uses this information to find out which meaning of "bat" is most appropriate. This type of analysis is crucial to various NLP applications, including machine translation, sentiment analysis, and information retrieval. In this type of task, precise understanding of word meanings can enhance the performance and reliability of NLP-based computer programs. The WSD process (approximated in Figure 1-4) [8] can be summarized as follows.

- **Context analysis**: Focus on the target word and analyze its surrounding context within the sentence.

- **Sense inventory**: For context analysis, generally, a predefined inventory or dictionary is used that lists different meanings (senses) of the target word.

- **Feature extraction**: Extract relevant features such as neighboring words, part-of-speech tags, and various syntactic dependencies.

- **Disambiguation algorithm**: Apply appropriate machine learning algorithms to determine the most likely sense (of the target word).

- **Evaluation**: Determine accuracy using various classification metrics like precision, recall, and F1 score against annotated data.

- **Application**: Integrate WSD techniques into NLP tasks that include the likes of machine translation and sentiment analysis.

This process may seem unfamiliar right now, but don't worry. You can just read it once and proceed even if you don't understand fully. We have included a code-based case study later in this book.

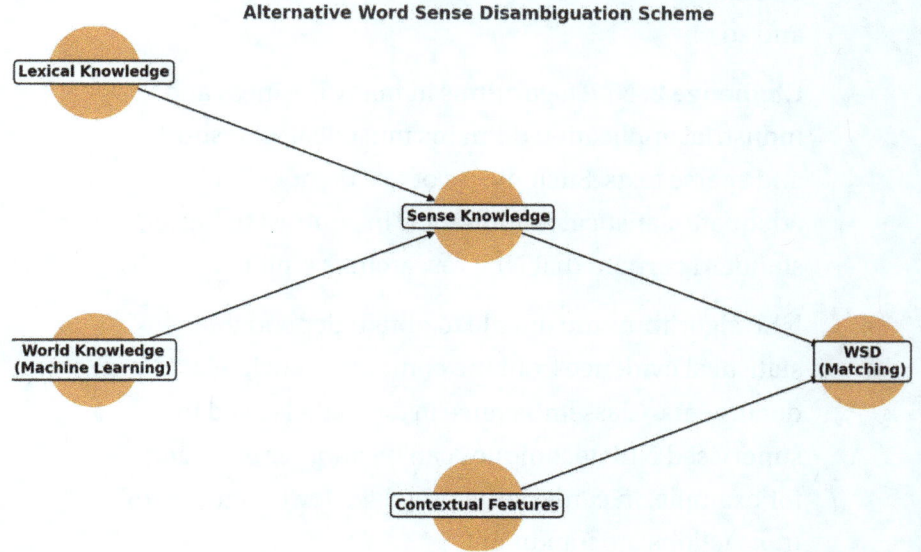

Figure 1-4. Workings of a WSD program

1.3 Challenges in NLP

NLP and its allied branches are continually growing in huge leaps and bounds with their ability to compute words and text. Given that, human language is unbelievably complex, fluid, and inconsistent. All natural languages, including English, present serious challenges that NLP researchers are continuously trying to overcome. A 2016 research paper presents a relatively complete picture of the major challenges. This paper divides these challenges into five items as follows.

- **Challenge 1**: It's listed as the variety and complexity in dealing with diverse and nonhomogeneous data sources. The text documents (in any text corpus) that we deal with are available in diverse forms and formats. This variety can come from many sources, such as documents encoded in different formats; the text length can also vary depending on the document. Documents can be available in multiple languages, and so on.

- **Challenge 2**: NLP algorithms in many business and industrial application domains must deal with short and sparse texts. Such pieces of text do not offer adequate statistical redundancy, in contrast to bigger, standard corpora that NLP researchers employ.

  NLP algorithms are unable to obtain dependable statistical evidence from the contents of such text documents. Class imbalance in data labels used in supervised NLP techniques can be another problem; for example, in card transaction data, fewer than 1% of transactions are fraudulent.

- **Challenge 3**: Evaluating and interpreting the results of NLP algorithms can be challenging due to the inherent subjectivity in certain business applications. For example, consider sentiment analysis of tweets, text classification, and text clustering. Gold-standard datasets used in academic research are rarely available in industrial and real-life situations.

  A related challenge is the lack of labeled data for training supervised machine learning modes for applications such as text categorization problems and sentiment analysis.

- **Challenge 4**: Human performance while dealing with natural languages is close to perfection. Given the current state of NLP research, computer programs cannot achieve this kind of 100% perfection. These limitations can be a challenge in certain situations, like dealing with space and aviation domains. A rising trend in many NLP applications, like text classification, is to use active learning to improve the NLP process and to get better results.

  By the way, active learning is a machine learning method where algorithms interactively query the user (or another source) to label new data points with the desired outputs. Imagine a computer program is learning to recognize different car models. The program might ask for human intervention when it identifies a specific car model in the photos, and it is not able to name it. These types of human or external agent interventions can improve the overall performance of the program.

- **Challenge 5**: While dealing with real-live and dynamic text data, as dealt with in certain business situations. This type of situation arises in applications that support decision-making in mission-critical events. In such cases, data containing text is required to be processed and get results in a timely and efficient way.

  A related challenge can be the rate at which data is generated in certain data streams. For example, in certain short messaging applications (tweets), around 6000 messages are generated every second on average. Accurately processing such high-paced text streams can be extremely challenging, as they demand a timely analysis to extract relevant and meaningful information from their contents.

These challenges can be viewed from multiple angles, and many more challenges can be identified for the NLP projects. This section serves the purpose of an introductory section. You will encounter many other challenges and their solutions as we progress through the book.

1.4 Applications of NLP

NLP applies to both text and speech, and it is equally applicable to all human languages—both spoken and written. So, there are numerous applications of NLP. Due to space limitations, we can only present a small fraction of them. Most of us interact with NLP almost on a daily basis, knowingly or otherwise. You might be familiar with applications like Siri personal assistant on spell checkers in word processors, iPhone, Alexa by Amazon, and many others. Some of these applications process human voice data and help resolve common queries. There are many other examples of NLP-powered tools that you may already be familiar with. Examples include email spam filters, AI-powered web search engines, machine translation of text or speech, automatic document summarization, social media sentiment analysis, and so on.

NLP applications are used by many industries. It has become an integral part of many industrial processes. NLP is revolutionizing the way customers interact with establishments; examples include NLP-driven customer service chatbots. Many business processes contain a substantial amount of unstructured text data, including social media messages, surveys, and emails. NLP-driven techniques can automate such processes with relative ease. The following are some examples of how NLP is helping business processes. Figure 1-5 depicts some interesting use cases of NLP.

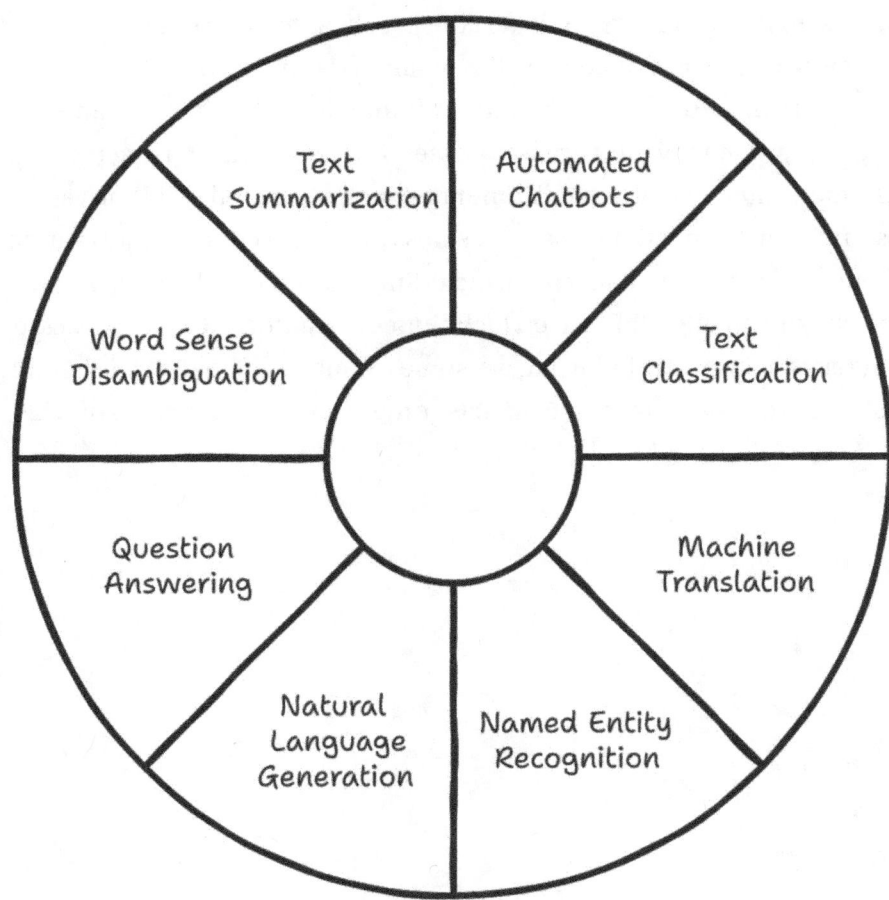

Figure 1-5. NLP use cases

1.4.1 Banking and Financial Industry (BFSI)

This industry widely utilizes NLP-driven chatbots, and digital virtual assistants are capable of providing instant and personalized services. Many banks and financial institutes are using NLP-driven financial fraud detection systems. Such systems immediately alert the back office staff when they detect any suspicious activity, such as fraudulent high-value credit card transactions. NLP can analyze a vast amount of text documents and extract only the relevant information necessary for maintaining regulatory compliance in industries such as pharmaceuticals and banking. NLP engines can extract insights from documents like financial reports, stock market data, and news articles that enable the industry to analyze risk and enable better decision-making. Many technologies-driven financial institutions are utilizing NLP in swift decision-making (and automation) in areas like loan and credit card applications, insurance claim documents, and contracts.

1.4.2 Healthcare

This industry encounters a large amount of unstructured text data, including medical records and other regulatory documents. Many healthcare companies are using NLP-driven engines to analyze the vast amount of information available in such documents to gain useful insights and to make data-driven decisions at all levels of the management hierarchy. The industry is utilizing clinical chatbots that assist doctors in diagnosing diseases. NLP analyzes patient feedback, online health forum data, and related social media messages to monitor patient sentiment and identify emerging health trends in community healthcare settings. This type of analysis is sometimes conducted in real-time, which can provide timely alerts for public health responses. This kind of swift response from the concerned professionals improves patient satisfaction as their concerns get addressed promptly.

1.4.3 Legal

Law professionals spend a large portion of their time searching for relevant information in the heap of a large collection of legal documents. This kind of legal research data can be a deciding factor in litigations, be it criminal or civil. NLP engines can automate a large portion of this legal discovery process, and legal professionals can spare more time to focus on another critical part of their cases. NLP can do document categorization by analyzing a vast amount of documents. This way, legal professionals can efficiently manage and organize large volumes of documents, enabling quick retrieval of relevant information.

These were just a few examples. NLP has applications in many other areas. We discuss a few next.

1.4.4 Automate Routine Business Processes

Chatbots and digital personal assistants, such as Siri from Apple and Alexa from Amazon, can recognize queries spoken in the form of human voice. They can match it to the appropriate entries from corporate or government databases to fetch the relevant information for users. After gathering the necessary information, digital assistants can formulate a response that is an appropriate answer to the user's query. In business operations, NLP automates data entry and extraction from word processors, email applications, and spreadsheets by processing and understanding unstructured data from emails, reports, and forms. They can even prioritize tasks based on content analysis, ensuring timely and organized execution. NLP NLP-based applications can automate many other routine, repetitive tasks to help organizations improve accuracy, reduce costs, and allocate resources more effectively.

1.4.5 Improve Search

NLP-based applications can improve keyword searches from a document or the Internet. They can help in the retrieval of relevant information for FAQ by disambiguating word senses based on context. For example, the English word spelled as "bat" can have different meanings based on the given context. NLP has the capability to resolve this ambiguity, and it is capable of fetching the right meaning of a word(s) based on the context. This can be very useful in identifying and resolving customer queries in FAQ applications and chatbots. Additionally, NLP-powered search engines can handle voice search effectively. It allows users to interact with search systems through spoken language to further enhance accessibility and convenience. For search engines, NLP can analyze user behavior and preferences based on past searches, nature of interactions, and demographics to personalize search recommendations. This customization capability helps in predicting user intent and delivering custom-tailored results that align with individual preferences and interests.

1.4.6 Search Engine Optimization

Search engine optimization (SEO) is the art of optimizing websites and their content to advance a website's visibility and ranking in search engine results pages (SERPs). The SEO process involves guessing which words people use the most when they search for the content of their interest (keyword research). It makes sure the content on websites is set up well for search engines, like Google Search, to understand. It is called on-page optimization. Another process, off-page optimization in SEO, involves activities outside the website to enhance the site's authority and credibility by increasing online mentions and search engine rankings.

Overall, SEO helps websites to appear higher in search results, which makes it more likely for people to click on them. NLP is a great tool in the entire SEO process as it can understand and analyze the intent behind user queries. Integrating NLP in SEO helps businesses cater to user needs more accurately and ultimately helps boost search engine rankings and drive more organic traffic to websites. To understand it better, organic traffic is the visitors who come to a website through unpaid search engine results, which excludes paid advertisements.

CHAPTER 1   NATURAL LANGUAGE PROCESSING: AN INTRODUCTION

1.4.7 Machine Translation: Translating Languages Automatically Using NLP

This technology has been under development till recently. Now, it has become quite sophisticated. NLP applications can produce quite accurate translations over a wide variety of natural languages. NLP is capable of processing both text and speech data and translating it accurately from one language to the other. Such machine translations are grammatically correct, and they can properly convey the meaning in the proper context.

Figure 1-6 illustrates a simplified NLP model.

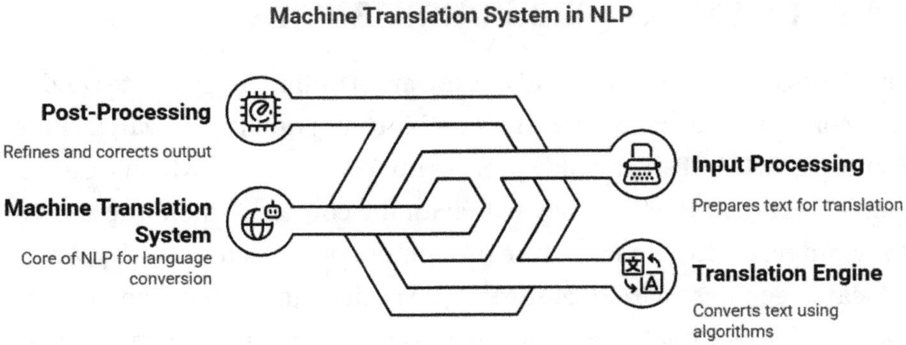

Machine Translation System in NLP

Post-Processing
Refines and corrects output

Machine Translation System
Core of NLP for language conversion

Input Processing
Prepares text for translation

Translation Engine
Converts text using algorithms

***Figure 1-6.** Machine translation model*

To a large extent, state-of-the-art NLP technology has overcome challenges such as handling ambiguous meanings of words in different contexts, preserving the original document's style and tone, and preserving the exactness of specialized domains. Machine translation is still an evolving field application of NLP.

1.4.8 Text Summarization: Condensing Large Text Information

A variety of test condensing definitions put the extent of summarization from 10% to 50% (see Figure 1-7). Text summarization applications use deep learning architectures, more specifically transformers. We discuss a few deep learning NLP techniques later in this book, but a detailed treatment of transformers is out of the scope. Extractive summarization techniques extract unchanged sentences from the original text corpus. These techniques are widely used by news aggregators, legal document reviewers, and academic researchers. The aim is to swiftly convey the important points of lengthy text documents to save readers time and effort.

Another category is LLMs like ChatGPT, which can generate original summaries using their own sentences (innovative) and are not available in the original text documents. They also use complex deep learning architectures based on transformer models. LLMs are very recent innovations that came up after 2020. Currently, such LLMs, though useful, can sometimes form some false innovative text on their own, a phenomenon known as hallucination. There is very serious research going on in the LLM domain, and their capabilities are improving with every new model release from the likes of Google and OpenAI.

CHAPTER 1   NATURAL LANGUAGE PROCESSING: AN INTRODUCTION

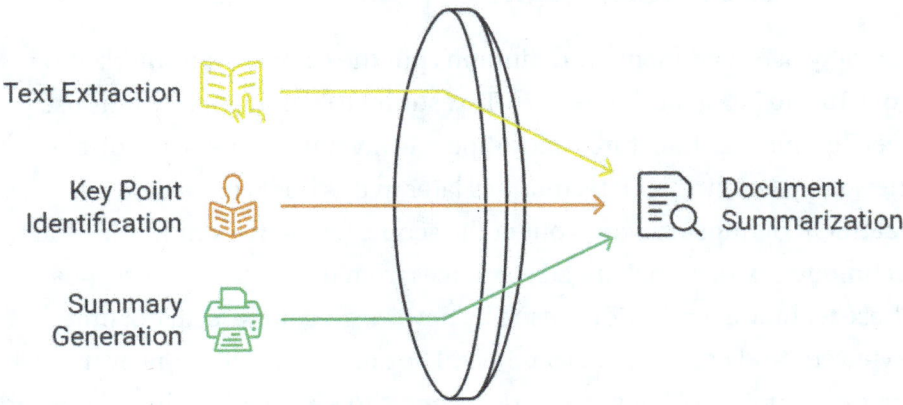

Figure 1-7. *Document summarization*

1.5 Recent Trends and Future Directions

NLP is a constantly evolving field. Many researchers from the domains of computer science and engineering, artificial intelligence and machine learning, and linguistics have actively contributed to the NLP. The main factors that have contributed immensely to the rapid developments in NLP are innovations in the fields of machine learning and an increase in computing power. Many recent trends seen in the NLP domain are made possible by deep learning and transformer models like BERT and GPT. These models have considerably improved the NLP capabilities of machines, particularly language understanding, translation, and generation. In recent years, the focus of NLP research has shifted to more context-aware and human-like communications. These developments are finding applications in enhancing the capabilities of chatbots and personal digital assistants, making them interact like humans. Other NLP areas that are gaining momentum are explainable NLP, bias mitigation, and ethical considerations of NLP programs.

Generally, the training of NLP models can consume a considerable amount of resources, requiring a huge text corpus for training and other software and hardware computing resources. The efforts are for developing low-resource language processing capabilities. Another focus area is automatic multilingual language processing. These recent developments are shaping to improve the way machines comprehend and interact with human language. The following are some examples of notable NLP trends.

1.5.1 Transfer Learning: Advantages of Pre-Training

Pre-trained models (or transfer learning models) are deep learning models that are trained on a huge corpus of data, which may be the data available on the entire accessible Internet. In other words, the models have learning stored in them in the form of deep learning parameters of weights, biases, and maybe other tunable parameters. This learning can be transferred to the other use cases as well by simply changing the last couple of layers in the original neural network architecture. This arrangement saves the model training time and the training data for the new uses. It's a huge saving on the overall resources as such. This phenomenon is popularly called transfer learning. They are t complex pre-trained deep learning models with uses in multiple use cases.

Transfer learning has revolutionized the NLP domain. They allow the model developers to transfer the pre-learned knowledge to other tasks, even if the tasks differ from those at the original training time. BERT, ELMo, GPT, ERNIE, ELECTRA, and RoBERTa are some of the popular pre-trained NLP models. These models were pre-trained on large corpora of text and other data and can be fine-tuned for other, more specific tasks. These models save huge time and resources for NLP programmers as they can eliminate the need to develop and train the NLP model code from scratch. Such pre-trained models require less labeled data for training while hugely

CHAPTER 1   NATURAL LANGUAGE PROCESSING: AN INTRODUCTION

saving on computing resources like memory, storage, and CPU power. These pre-trained models are attractive options for developers for specific tasks like language translation, chatbots, social media sentiment analysis, topic modeling, and text summarization. Figure 1-8 compares transfer learning with the traditional approach.

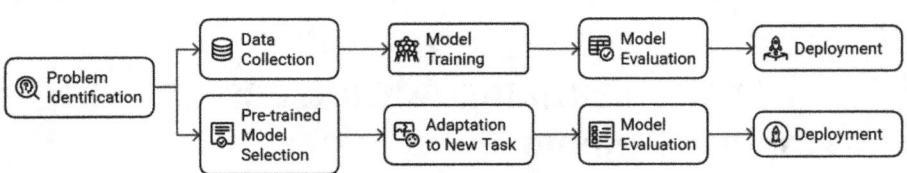

Figure 1-8. *Traditional machine learning and pre-trained models*

Wang et al. (2022) have given a complete review of the methods and frameworks of transfer learning models. They first give a short introduction to transfer learning models, which is followed by representative methods and applicable frameworks. This paper then goes on to analyze the effects and challenges of transfer learning models. Finally, the paper briefly concludes by reporting future research trends in the domain of pre-trained or (transfer learning models).

1.5.2 LLMs: Human-like Interactions

A large number of research papers are available that discuss the applications of LLMs, like ChatGPT, in education, medicine, and many other domains. Not surprisingly, almost everyone around knows about ChatGPT, and there is a large probability that they might have even used it for their benefit. Enkelejda et al. discuss specifically the current state of ChatGPT (and related software) and its applications in the education domain. They discuss how ChatGPT can be used to develop quality educational content that can improve student engagement and interaction and provide more

CHAPTER 1   NATURAL LANGUAGE PROCESSING: AN INTRODUCTION

personalized learning experiences. The paper argues that, before using it, students and teachers must understand about the limitations and (at times) unanticipated brittleness of such systems. Continuous human oversight is necessary to deal with the expected bias of output from such LLMs. The paper finally concludes with recommendations for how to resolve the common challenges posed by LLMs.

By definition, LLMs are complex deep learning models that can understand and generate human language text. They can interact just like humans and can talk on almost any subject under the sun, though they can make serious mistakes at times (called hallucination). LLMs are trained on huge datasets, so the name is large. The size of LLM training datasets can range from a thousand to a million gigabytes or even larger. LLMs can be further trained via tuning: they can be fine-tuned to a particular task or domain. For example, LLMs can be fine-tuned to perform finance, healthcare, or education-related tasks. LLMs can be used in almost every domain, including chatbots, sentiment analysis, customer service, DNA research, and online search.

Some examples of popular LLMs include GitHub's Copilot, OpenAI's ChatGPT, Bard from Google, Llama from Meta, and Microsoft's Bing Chat. More specific tasks that these LLMs can perform include text generation, summarization, translation, question answering, code generation (Copilot), and sentiment analysis.

1.5.3 Cross-Lingual and Multiple-Lingual Models: Handling Multiple Languages

Cross-lingual NLP models can comprehend multiple languages, and they are also capable of transferring knowledge between them. This can swiftly enable tasks like machine translation and cross-lingual understanding. Although a multiple-lingual model (MLM) models can also handle multiple languages simultaneously, this understanding comes without necessarily transferring knowledge between multiple languages. MLMs can perform many different NLP tasks without the need to have separate language models for each language. A single MLM can handle multiple languages parallelly.

The tasks that can be performed by cross-lingual NLP models include machine translation, social media sentiment analysis, cross-lingual information retrieval, multilingual text classification, cross-lingual question answering, entity recognition, and language inference. At the same time, MLMs are useful in tasks like text classification, language detection, named entity recognition, part-of-speech tagging, sentiment analysis, machine translation, question answering, and text summarization.

The real-world examples of multilingual NLP models (MLMs) include Google Translate, Microsoft Translator, Amazon Comprehend, Facebook M2M-100, and IBM Watson Language Translator. In contrast, cross-lingual NLP models are represented by XLM-R (Cross-lingual Language Model–RoBERTa), mBERT (multilingual BERT), M2M-100 (Many-to-Many 100), LASER (Language-Agnostic SEntence Representations), and T5 (Text-To-Text Transfer Transformer).

1.5.4 Explainable NLP: Reason Model Results

In simple words, explainable NLP is like having a colleague who explains why she thinks something is true so that you can understand her reasoning. As put forward by Søgaard (2021) in his NLP book, explainable NLP is an emerging field that is lately getting a lot of attention from the tech community worldwide. Many industries, including healthcare and law, require the NLP developers to reason the predictions made by the models developed by them. In fact, it's a legal requirement in many countries.

The knowledge of explainable NLP is essential for two reasons: one is confirming that the AI-driven solutions duly comply with local and international regulations, particularly in finance and other sensitive fields like healthcare and legal. Another reason is that if we better understand how a model works, it can translate to reduce errors and develop capabilities anticipating the strengths and weaknesses of the model. Most importantly, a deep understanding of the inner workings of NLP models (explainable NLP) can help in avoiding unexpected behaviors of the NLP models that are already in production. This knowledge can also help in eliminating the impact of social biases, shown by many NLP and other machine learning models. Bias can appear in models because of the use of biased training data. Thus, the knowledge of explainable NLP translates into enhanced trust and confidence when deploying an NLP-powered solution in production.

1.5.5 Emergent Areas in NLP

The boundaries of NLP research are expanding with every passing day. Emotional AI is emerging, where NLP models can analyze text and speech patterns to realize human emotions. Developments in multilingual processing using LLMs like ChatGPT are defying language barriers and promoting global communication. The NLP models are getting integrated with other modalities that include images and speech (Multimodal NLP). NLP is getting integrated with augmented reality, which allows different NLP technologies to interact within immersive environments. With real-time and interactive NLP, the speed and responsiveness of NLP models will progress. This improves the capabilities and response time in applications like real-time language translation (like those in international VIP conversations), video captioning, and Alexa-like interactive voice assistants.

Ethical and responsible AI is another research area that is getting considerable attention. Researchers are constantly working toward addressing biases, ensuring data privacy, and maintaining transparency in AI systems. Future NLP models, including those in the GenAI domain, are expected to show better generalization capabilities. They are more reliable, requiring less fine-tuning data to adjust to new tasks. In the coming years, we can expect even larger, more powerful, and more reliable language models (LLMs included). These models have a sophisticated understanding of context, which allows them to produce more precise and contextually appropriate responses.

1.6 Short Business Case 1: Application of NLP in Medical Text Analysis

Around the globe, a vast amount of text data is generated by the healthcare industry in the form of including clinical notes, patient records, research articles, and much more. Extracting actionable insights from such text data is a challenging but essential task. These valuable insights can be helpful in improving patient care, advancing medical research, and streamlining operations. Many powerful NLP-based solutions exist today to enable the analysis and interpretation of medical texts in different languages—transforming unstructured data into valuable information.

NLP applications can considerably add value to the patient healthcare business by extraction of critical information from electronic health records. For example, NLP can help by identifying patterns in symptoms, diagnoses, and treatments. This can ultimately help in early-stage disease diagnosis and more customized treatment plans. An NLP-assisted system could give an early warning for a potential diagnosis of diabetes in patients with specific symptom patterns and lab results. It can lead to the instigation of further and timely intervention by medical professionals.

1.6.1 Streamlining Clinical Research

Medical research professionals often dig through large piles of electronic and other health records (relevant research, patient data, and results of clinical trials) for their investigations. NLP applications can automate the extraction of relevant information from research articles, clinical

CHAPTER 1   NATURAL LANGUAGE PROCESSING: AN INTRODUCTION

trial databases, and patient records. For example, these applications can quickly identify studies that match their criteria or find patients with specific genetic markers for targeted trials.

1.6.2 Improving Administrative Efficiency

Many administrative tasks, like coding, billing, and documentation, can be very laborious, time-consuming, and prone to errors at the same time. NLP applications can streamline these business processes by automating the extraction and classification of information needed for billing codes, patient insurance claims, and essential regulatory compliance. For example, NLP systems are capable of reading and interpreting doctors' notes and other medical records to automatically extract billing codes. This can considerably reduce the administrative burden on healthcare providers and, at the same time, minimize the risk of errors.

1.6.3 Enhancing Data Accessibility

Manual storage and retrieval of patient and other health records is a very laborious and time-consuming task. NLP applications can automate the entire process. In fact, most healthcare establishments these days are already use these applications. By converting unstructured text records into structured data, NLP applications are capable of creating searchable databases of medical records. This enables quick access to patient histories and treatment outcomes, enabling quick decision-making and coordination among healthcare professionals. Additionally, NLP can also be used to create patient-friendly summaries (in their mother tongues, if required) of complex medical information. Such summaries in local languages can improve patient understanding and engagement in their care.

31

1.6.4 Facilitating Predictive Analytics

NLP applications can make accurate data analysis to forecast patient outcomes and identify potential risks. NLP can analyze relevant data from unstructured medical texts and enhance predictive analytics in healthcare applications. For instance, by analyzing historical electronic health records, NLP applications can identify risk factors for readmission or complications. This can enable proactive interventions in patient treatments, improving patient outcomes and possibly reducing healthcare costs.

1.6.5 Business Case Conclusion

Along with these resource-saving applications of NLP applications in healthcare, there are many challenges like complexity and variability of medical records or language. Ensuring the accuracy and reliability of NLP applications is critical. It's especially true in a medical context where even small errors can have serious consequences. No analysis in the world of AI is 1005 accurate—NLP included. Continuous refinement and development of more efficient algorithms and collaboration with medical professionals are necessary to further improve the precision and applicability of NLP in healthcare.

Medical NLP applications hold incredible potential for renovating healthcare by enhancing patient care, streamlining clinical research, improving administrative efficiency, and facilitating predictive analytics. By leveraging NLP technologies, healthcare professionals can unlock valuable insights from unstructured data, which can lead to better patient outcomes and more efficient healthcare systems. As NLP in healthcare continues to evolve, in the future, it can play a pivotal role in advancing personalized medicine and optimizing healthcare delivery.

Many more business cases can be listed that utilize NLP for applications in other domains like healthcare, finance, legal, e-commerce, education, customer service, social media, marketing, news media, human resources, travel, and entertainment. The possibilities with NLP are endless. Figure 1-9 depicts a few more use cases of NLP-based applications.

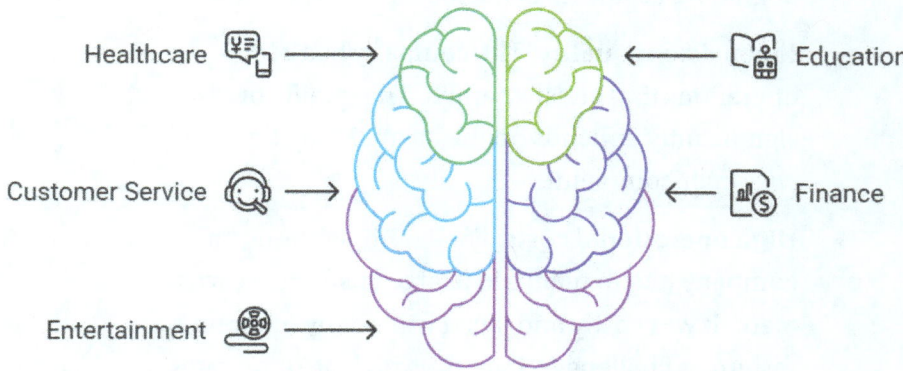

Figure 1-9. More uses of NLP leveraged technologies

1.7 Short Business Case 2: NLP in Customer Service, Enhancing Customer Interaction

ABC Inc. is a leading apparel e-commerce company based in downtown New York. It is known for its varied product range and customer-first approach. In a couple of decades, as the company grows, the volume of online customer interactions also increases. Maintaining high levels of customer satisfaction becomes more important with every passing day. To solve this and some other related challenges, the company decided to implement a custom-made NLP application that can help enhance its customer service operations and maintain customer satisfaction levels.

1.7.1 The Challenge

Prior to the implementation of the NLP solution, ABC Inc. was challenged with several issues in its customer service department.

- **High volume of inquiries**: The client service team faced many issues with managing the daily large number of inquiries. It led to delayed responses and degrading customer dissatisfaction.

- **Inconsistent quality**: The company has 24/7 operations that are all manual. The quality of responses significantly depends on the agent. Many times, it led to customer complaints.

- **High operational costs**: For 24/7 operations, the company had to maintain a large customer service team. It was costly and faced with many other human resources challenges, including high attrition rates.

- **Lack of insights**: ABC Inc. was not able to analyze the customer sentiments and common issues based on the voice and other text data it had. It added to existing operational inefficiencies. Implementing proactive measures was one of the top priorities for the company.

1.7.2 Implementation of NLP

To overcome these challenges, ABC Inc. decided to leverage multiple NLP technologies, including automated chatbots, sentiment analysis from the customer inquiries text data, classification of text and voice inquiries, and natural language understanding (NLU).

CHAPTER 1   NATURAL LANGUAGE PROCESSING: AN INTRODUCTION

ABC Inc. introduced AI-powered chatbots that were capable of handling routine inquiries efficiently. Instant responses were provided to many of the routine inquiries. The training of these chatbots was done using historical customer service data to make it easier to understand and respond to common questions about orders, returns, and product information.

Customer sentiment analysis was done from customer feedback data from various channels, including emails, social media, and online reviews. This helped the company judge customer sentiment in real time and address issues promptly. Text classification solutions help to automatically classify incoming customer inquiries. This classification was done based on the content of inquiries. This classification helped in efficient routing to the appropriate department or escalation to human agents (manages if required) if necessary.

The newly implemented chatbots and virtual assistants were equipped with advanced NLU capabilities. They were capable of understanding the context and nuances of customer inquiries. This automation led to more accurate and helpful responses.

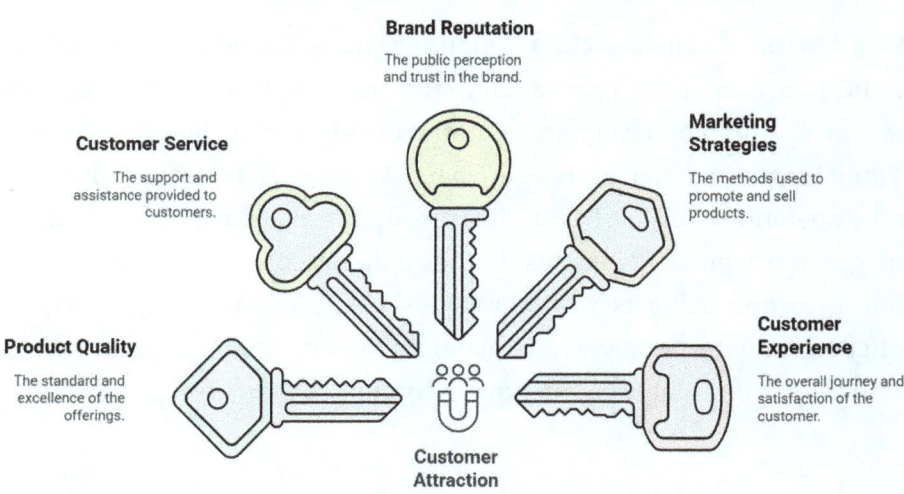

Figure 1-10. Critical elements that attract today's customers

1.7.3 Results and Benefits

The implementation of NLP at ABC Inc. resulted in several significant benefits that included enhanced customer satisfaction levels, improved response time,

Improved Response Times: With the introduction of chatbots, ABC Inc. was able to provide instant responses to a large number of inquiries, significantly reducing wait times for customers, generating actionable insights, scalability of operations, and reducing costs. The company saved money by reducing human agents (for 24/7 operations) as chatbots started doing many inquiry-handling jobs. The generation of actionable insights into customer perceptions and common pain points enabled the company to proactively address issues and improve its products and services based on customer feedback. A high level of automation at the operational level made the scalability of the business possible with reduced costs. Now, during peak hours, almost 80% of inquiries were getting answered without the need for any human intervention.

1.7.4 Conclusion

The adoption of multiple NLP solutions has transformed customer service at ABC Inc. Automation is enhancing customer interaction and operational efficiency. Automated chatbots, sentiment analysis, text classification, and natural language understanding are being leveraged to increase efficiency at the operational level. ABC Inc. has mostly addressed the key challenges and secured significant benefits. As the company continues to grow, it plans to further refine its NLP capabilities. They plan to provide even better customer experiences and maintain their competitive edge at the same time in this highly competitive e-commerce industry.

CHAPTER 1   NATURAL LANGUAGE PROCESSING: AN INTRODUCTION

1.8 Example with Python

This section presents an NLP getting-started tutorial with an explanatory commentary. The Python code used in this tutorial is also well-commented. The business problem is to predict whether the test tweets are positive or negative. The training data is provided with proper labels. And the test tweets have no labels. Our job is to write Python code to predict the test data tweets and to evaluate the F1 score of our model. We will be using a random forest algorithm to make our predictions.

1.8.1 NLP Tutorial
1.8.2 YouTube Comments Spam Detection Using Python

The first NLP tutorial uses the YouTube Spam Collection Dataset. We use a simple technique to convert the text data into a numeric format that can be processed by machine learning algorithms. Next, we apply the random forests algorithm to build a machine learning model. Finally, we make the predictions for the target variables and also evaluate our model's accuracy. This book expects you to have some prior knowledge of Python and coding machine learning and deep learning (neural networks) models. Let's jump into writing our code and try to understand it as we write it.

```
# Import the required libraries.
import pandas as pd
import numpy as np
# The below code is for working with machine learning model.
from sklearn import feature_extraction, linear_model, model_selection, preprocessing
from sklearn.model_selection import train_test_split
from sklearn.ensemble import RandomForestClassifier
```

CHAPTER 1 NATURAL LANGUAGE PROCESSING: AN INTRODUCTION

```python
from sklearn.metrics import accuracy_score
# Ignore warnings.
import warnings
warnings.filterwarnings('ignore')
```

The Data

Let's look at our data.

```python
# Read the data files available in the same folder as this code.
Youtube01_psy = pd.read_csv('Youtube01-Psy.csv')
Youtube02_katyperry = pd.read_csv('Youtube02-KatyPerry.csv')
Youtube03_lmfao = pd.read_csv('Youtube03-LMFAO.csv')
Youtube04_eminem = pd.read_csv('Youtube04-Eminem.csv')
Youtube05_shakira = pd.read_csv('Youtube05-Shakira.csv')
# Let's check the datasets size.
print(Youtube01_psy.shape)
print(Youtube02_katyperry.shape)
print(Youtube03_lmfao.shape)
print(Youtube04_eminem.shape)
print(Youtube05_shakira.shape)
(350, 5)
(350, 5)
(438, 5)
(448, 5)
(370, 5)
# ACombine all five datasets.
combined_df = pd.concat([Youtube01_psy, Youtube02_katyperry,
Youtube03_lmfao, Youtube04_eminem, Youtube05_shakira])

# Reset the index
combined_df.reset_index(drop=True, inplace=True)
combined_df.head(3)
```

CHAPTER 1   NATURAL LANGUAGE PROCESSING: AN INTRODUCTION

```
                                 COMMENT_ID            AUTHOR  \
0   LZQPQhLyRh8oUYxNuaDWhIGQYNQ96IuCg-AYWqNPjpU       Julius NM
1   LZQPQhLyRh_C2cTtd9MvFRJedxydaVW-2sNg5Diuo4A      adam riyati
2   LZQPQhLyRh9MSZYnf8djykOgEF9BHDPYrrK-qCczIY8  Evgeny Murashkin

                  DATE                                   CONTENT  \
0  2013-11-07T06:20:48   Huh, anyway check out this you[tube]
channel: ...
1  2013-11-07T12:37:15   Hey guys check out my new channel and
our firs...
2  2013-11-08T17:34:21    just for test I have to say murdev.com

   CLASS
0      1
1      1
2      1
# Select only the useful "CONTENT" and "CLASS" columns.
combined_df = combined_df[["CONTENT", "CLASS"]]
# Randomly select 5 rows
random_sample = combined_df.sample(n=5)
print(random_sample)
                                                CONTENT  CLASS
987                          subscribe to my chanell         1
1727                        BEST SONG! GO SHAKI :D          0
1280   Share Eminem&#39;s Artist of the Year video so...  1
1241   MEGAN FOX AND EMINEM TOGETHER IN A VIDEO  DOES...  0
1126             I learned the shuffle because of them    0
```

Note the emojis and URLs in CONTENT and the misalignment due to spaces in CLASS.

```
combined_df.shape
(1956, 2)
# Map "0" to  "Not Spam" and 1 : "Spam."
```

CHAPTER 1   NATURAL LANGUAGE PROCESSING: AN INTRODUCTION

```
#combined_df['CLASS'] = combined_df['CLASS'].map({0 : "Not
Spam", 1 : "Spam"})
# Randomly select 5 rows
random_sample = combined_df.sample(n=5)
print(random_sample)
                                           CONTENT   CLASS
1003               Check out this playlist on YouTube:    1
1455                              So freaking sad...      0
474   Imagine this in the news crazy woman found act...   0
1343  I know that maybe no one will read this but PL...   1
37    SUB 4 SUB PLEASE LIKE THIS COMMENT I WANT A SU...   1
# Seperate features and the target.
X = np.array(combined_df['CONTENT'])
y = np.array(combined_df['CLASS'])
X.shape
(1956,)
```

Building Vectors

The words contained in every tweet are a good pointer of whether they are about a real disaster or not. In theory, this is not totally correct. We will still use it as our starting point in our first NLP tutorial.

The following uses scikit-learn's CountVectorizer to count the words in each tweet and then turn them into a data format that our machine learning model can understand. A vector is, in this context, a set of numbers that a machine learning algorithm can understand. The example shows the related code, its usage, and output.

```
demo_text = ["Stella is a good girl. She loves to swim"] # Demo
sentence
count_vectorizer = feature_extraction.text.CountVectorizer()
# Instrantiate CountVectorizer()
count_vectorizer.fit(demo_text) # Fit the demo text
```

```
print(count_vectorizer.vocabulary_) # Print results
{'stella': 5, 'is': 2, 'good': 1, 'girl': 0, 'she': 4,
'loves': 3, 'to': 7, 'swim': 6}
# encode document
demo_vector = count_vectorizer.transform(demo_text)
print(demo_vector.shape)
print(demo_vector.toarray())
(1, 8)
[[1 1 1 1 1 1 1 1]]
```

There are eight number elements in the vector representing demo_text [[1 1 1 1 1 1 1 1]] (0 to 7) and the number of distinct words is also eight.

All the entries in the vector are 1 as no word in the demo_text is repeating (frequency of all words is 1).

For more information on CountVectorizer in Python, see `https://www.educative.io/answers/countvectorizer-in-python`.

A Second Example

Let's try another example.

```
demo_text2 = ["The sky is blue. I wish to fly in the blue sky"]
# Demo sentence 2
count_vectorizer = feature_extraction.text.CountVectorizer()
# Instrantiate CountVectorizer()
count_vectorizer.fit(demo_text2) # Fit the demo text
print(count_vectorizer.vocabulary_) # Print results
{'the': 5, 'sky': 4, 'is': 3, 'blue': 0, 'wish': 7, 'to': 6,
'fly': 1, 'in': 2}
# encode document
demo_vector = count_vectorizer.transform(demo_text2)
```

```
print(demo_vector.shape)
print(demo_vector.toarray())
(1, 8)
[[2 1 1 1 2 2 1 1]]
```

We are simply counting the repetition of words and putting it in the vector. The words "the", "sky", and "blue" have a frequency of two each in the demo_text2. So, these three words are represented by two each in the demo vector [[2 1 1 1 2 2 1 1]].

Counts for the First Five Entries

Let's get counts for the first five entries in combined_df.

```
## let's get counts for the first 5 entries in
"CONTENT" column.
demo_vectors = count_vectorizer.fit_transform(combined_
df['CONTENT'][0:5])
type(demo_vectors) # Checking data type.
scipy.sparse._csr.csr_matrix
demo_vectors.shape # Checking shape.
(5, 46)
# we use .todense() here because these vectors are "sparse"
# (only non-zero elements are kept to save space)
print(demo_vectors[0].todense().shape)
print(demo_vectors[0].todense())
(1, 46)
[[0 1 0 1 1 0 0 0 0 0 0 0 1 0 0 0 1 0 0 0 0 0 0 0 0 1 0 0
  0 0 0 0 0 1 0 1 0 0 0 0 0 1]]
```

The preceding code and result show the following.

- There are 46 distinct words (called "tokens") in the selected first five tweets.

CHAPTER 1   NATURAL LANGUAGE PROCESSING: AN INTRODUCTION

- Obviously, the first tweet (and every other tweet) contains only some of those 46 distinct tokens.

- The vector contains 54 elements because there are 54 distinct tokens.

- All of the non-zero counts in the vector are the tokens that definitely exist in the entry in the "CONTENT" column.

Now, let's create representative number vectors for all of the five entries.

```
combined_df.shape
(1956, 2)
X.shape
(1956,)
# Split in to train and test datasets.
X_train, X_test, y_train, y_test = train_test_split(X, y,
test_size=0.3, random_state=42)
X_train.shape
(1369,)
X_test.shape
(587,)
# We will use count_vectorizer.fit_transform() for X_train
and X_test.
X_train = count_vectorizer.fit_transform(X_train)
X_test = count_vectorizer.transform(X_test) # We will do onlt
transform() with X_test.
X_train.shape
(1369, 3458)
X_test.shape
(587, 3458)
```

43

CHAPTER 1　NATURAL LANGUAGE PROCESSING: AN INTRODUCTION

Let's build a simple machine learning model to predict the "target" variable.

```
# Initialize the Random Forest model
clf = RandomForestClassifier(n_estimators=100, random_state=42)
```

Next, let's train the model and evaluate it. The evaluation metric used for this is an F1 score.

```
combined_df.columns
Index(['CONTENT', 'CLASS'], dtype='object')
scores = model_selection.cross_val_score(clf, X_train, y_train,
cv=3, scoring="f1")
scores
array([0.95067265, 0.95594714, 0.95594714])
```

The F1 scores look good! You can get better results with other NLP techniques that we cover in upcoming chapters.

Next, let's make the predictions on the test data.

```
X_train.shape
(1369, 3458)
clf.fit(X_train, y_train)
RandomForestClassifier(random_state=42)
X_test.shape
(587, 3458)
# Make predictions
y_pred = clf.predict(X_test)
y_pred.shape
(587,)
# Construct a dataframe with columns as y_test and y_pred.
test_df = pd.DataFrame()
test_df["y"] = y_test
test_df["y_predict"] = y_pred
```

CHAPTER 1 NATURAL LANGUAGE PROCESSING: AN INTRODUCTION

```
# Display 10 random rows from test_df
random_sample = test_df.sample(n=10)
print(random_sample)
     y  y_predict
454  1          1
475  0          0
252  1          1
298  1          1
108  1          1
316  1          1
101  1          1
230  0          0
437  1          1
428  0          0
```

We are finishing the solution here, though the predictions also look good.

For real-world analysis, the F1 score should be calculated on the test data. We are skipping this step.

Note We skipped much of the data cleaning and preprocessing in this tutorial, but we managed to get respectable model results. Chapter 5 takes up the same tutorial again, but this time with reasonable cleaning of data. It will be interesting to see if there are any further improvements in the model results with data cleaning done.

This completes our job for now. Remember, it was an oversimplified example created only for the demo.

1.9 Recap of Key Concepts

This chapter gives a thorough exploration of NLP and its implications in today's industrial landscape. This chapter begins by establishing the importance of NLP, followed by a deep dive into fundamental techniques utilized today in the NLP domain. Then, our discussion turns to the key NLP concepts, including text classification, clustering, and collocation. Other essential processes like computing word frequencies, text extraction, WSD, and stop word removal are also discussed thereafter. It's explained how NLP applications resolve the confusing meanings of a word in different contexts. The chapter also delves into the challenges faced by NLP.

The text then proceeds to explore the real-life applications of NLP across industrial domains, including banking and financial services, the use of NLP for analyzing financial documents, and automating customer interactions through chatbots and personal digital assistants. In the world of medicine and healthcare, NLP finds applications in text analysis and improving patient care through enhanced data insights. The legal community uses NLP for document reviews and legal research, text summarization, and to automate other routine business processes. The chapter also highlights recent trends and future directions in NLP, such as transfer learning, LLMs, and cross-lingual models. These relatively recent NLP applications are powering the development of more sophisticated and human-like interactions in NLP-based systems.

Finally, we have included two short business cases on applications of NLP in the industry. The first case focuses on medical text analysis. It demonstrates how NLP tools can enhance the extraction and interpretation of medical information. The second case examines the application of NLP in customer service and business process outsourcing. The chapter ends with a fully solved (using Python) business case that demonstrates how to apply NLP key concepts in analyzing tweets. The primary purpose of this tutorial is to get you started with Python coding in the NLP domain.

1.10 Conclusion

The overview of NLP presented in this chapter is enough to develop a comprehensive understanding of the crucial role the NLP plays in modern technology. This chapter helps in developing essential NLP techniques, such as text classification, clustering, and WSD. This understanding helps lay the groundwork for understanding how NLP processes analyze unstructured textual data. Discussing the challenges faced in NLP helps you understand the complexities involved in developing robust NLP systems that can handle linguistic nuances. The varied applications of NLP across many industry sectors, including banking and finance, healthcare, and legal, focus on its long-term transformative impact. NLP automates routine business processes like SEO, improves the search capabilities of modern search engines, automates cross-language translations, summarizes text, and helps to improve customer interactions. These applications further demonstrate the value of NLP in practical business contexts.

Looking forward, the chapter examines recent trends and future directions in NLP research (i.e., transfer learning, LLMs, and cross-lingual capabilities) to highlight the ongoing advancements in the field. Continued research in the NLP domain is critical for addressing evolving challenges and enhancing the effectiveness of current NLP systems—ongoing exploration and development in the NLP domain are essential for leveraging its full potential.

1.11 Exercises

1. We solved the tutorial problem in this chapter with the classification technique. Try to apply clustering algorithms (e.g., k-means) to arrange the text samples into clusters based on similarity.

      a. Can clustering help identify hidden patterns or get to any themes within the training dataset?

      b. Make a presentation: Two groups will present their classification and clustering results and explain their process and any challenges faced.

      c. Discussion points

           a. How did text classification and clustering help in organizing the data?

           b. What were the main challenges encountered during this exercise?

           c. How can these techniques be applied in real-world scenarios?

2. Read the two short business cases given in this chapter and prepare a small presentation to discuss the following in your class.

      a. What are the most common NLP techniques used across different industries?

      b. How does NLP enhance efficiency and effectiveness in each case study?

      c. What are the potential future developments in NLP for these industries?

3. Apply TF-IDF (term frequency-inverse document frequency) and apply it to the Python tutorial problem. Do you see any improvements in results? This technique is covered in the upcoming chapters.

CHAPTER 1   NATURAL LANGUAGE PROCESSING: AN INTRODUCTION

1.12 References

[1] Alberto, T. C. and Lochter, J. V. (2017). YouTube Spam Collection. UCI Machine Learning Repository. https://doi.org/10.24432/C58885.

[2] Amazon Web Services (2024). Available at: https://aws.amazon.com/what-is/nlp/. Last Accessed: July 6, 2024.

[3] IBM (2024). Available at: https://www.ibm.com/topics/natural-language-processing. Last Accessed: July 6, 2024.

[4] Donges, N. (2023). Introduction to Natural Language Processing (NLP), Available at: https://builtin.com/data-science/introduction-nlp. Last Accessed: July 6, 2024.

[5] Srinivasa-Desikan, B. (2018). *Natural Language Processing and Computational Linguistics: A Practical Guide to Text Analysis with Python, Gensim, spaCy, and Keras*. Packt Publishing Ltd.

[6] Duarte, F. (2024). Amount of Data Created Daily, Available at: https://explodingtopics.com/blog/data-generated-per-day. Last Accessed: July 6, 2024.

[7] El Naqa, I., and Murphy, M. J. (2015). *What is machine learning?* (pp. 3-11). Springer International Publishing.

[8] Suhail, A. (2023). *Text Multi-Classification with Flair NLP*. Available at: https://medium.com/tensor-labs/text-multi-classification-with-flair-nlp-4a8af200b7c. Last Accessed: July 9, 2024.

49

[9] Zhou, X., and Han, H. (May 2005). "Survey of Word Sense Disambiguation Approaches." In *FLAIRS* (pp. 307–313).

[10] Ittoo, A., and van den Bosch, A. (2016). "Text analytics in industry: Challenges, desiderata and trends." *Computers in Industry*, 78, 96–107.

[11] Text Summarization- An emerging NLP technique for your business requirements, Available at: https://medium.com/@annoberry/text-summarization-an-emerging-nlp-technique-for-your-business-requirements-daf5f560315d. Last Accessed: July 13, 2024.

[12] Geeksforgeeks (2024). Machine Translation in AI, Available at: https://www.geeksforgeeks.org/machine-translation-of-languages-in-artificial-intelligence/. Last Accessed: July 13, 2024.

[13] Annoberry Technology Solutions (2022). Text Summarization: An emerging NLP technique for your business requirements. Last Accessed: July 13, 2024.

[14] Becker, C., Hahn, N., He, B., Jabbar, H., Plesiak, M., Szabo, V., ... and Wagner, J. (2020). Modern Approaches in Natural Language Processing. LMU Munich, Munich.

[15] Wang, H., Li, J., Wu, H., Hovy, E., and Sun, Y. (2022). Pre-trained language models and their applications. Engineering.

[16] Kasneci, E., Seßler, K., Küchemann, S., Bannert, M., Dementieva, D., Fischer, F., and Kasneci, G. (2023). "ChatGPT for good? On opportunities and challenges of large language models for education." *Learning and Individual Differences*, 103, 102274.

[17] Thirunavukarasu, A. J., Ting, D. S. J., Elangovan, K., Gutierrez, L., Tan, T. F., and Ting, D. S. W. (2023). Large language models in medicine. Nature medicine, 29(8), 1930–1940.

[18] Søgaard, A. (2021). *Explainable natural language processing.* Morgan & Claypool Publishers.

[19] Ariwals, p (2024). *Top 14 Use Cases of Natural Language Processing in Healthcare.* Available at: https://marutitech.com/use-cases-of-natural-language-processing-in-healthcare/. Last Accessed: July 21, 2024.

[20] Khubwani, H. (2023). *Why Customer Experience Should be the Top Accessory in the Fashion Industry.* Available at: https://www.clootrack.com/blogs/customer-experience-in-fashion-industry. Last Accessed: July 21, 2024.

CHAPTER 2

Collecting and Extracting the Data for NLP Projects

2.1 Why You Should Learn NLP

There has been an explosion in the availability of data in every domain these days. The average amount of daily data generated globally is in the hundreds of exabytes. It is an enormous amount of data by any standard, with a substantial contribution coming from the United States. This data is useful to government and businesses only when it is analyzed properly and ultimately results in actionable items. In this data-driven world, natural language processing, or NLP, is a recommended tool that supports businesses across various domains in analyzing text-based, structured, and unstructured data.

The primary sources of text data available to businesses typically include interactions through social media, customer feedback, and numerous other similar sources. The proper analysis of data supports a better understanding of customers by analyzing market trends and finally making informed decisions. By implementing NLP-based text analytical techniques, companies can benefit from improved operational efficiency and significantly support their strategic decision-making processes.

The effective implementation of NLP in businesses can reduce manual work for employees and management and improve overall efficiency. The handling of a large volume of text data is completely automatic. The outcome of successful NLP implementation is the early detection of potential issues followed by the generation of the most appropriate action items required for dynamically changing markets. This reduces the overall reaction time, saving a significant amount of valuable time for management, which can be allocated to other critical business activities.

So, businesses can benefit from the predictive capabilities of NLP to forecast customer needs and related market shifts. The companies gain a competitive edge through the automatic generation of these valuable foresights and the resultant proactive strategy adjustments. So, it can be appreciated that NLP is a crucial tool that can generate actionable business insights that can help to run efficient operations and improve the company's bottom line in today's rapidly evolving marketplace.

Provide a case study or business example of how some well-known corporation achieved insights by collecting and processing NLP data.

2.2 Where to Find the Data (Text Corpora) for NLP Projects

Real-time business projects in NLP typically utilize vast amounts of linguistic data. Such data is called *text corpora* in the NLP community. There is another related term: *lexical resources*. Let's discuss what it is and how it is related to the term text corpora.

Text corpora are large collections of written texts. Many corpora are designed to comprise a thoughtful mix of content across different categories. Such special corpora are used to study how different specialized societies use language. Lexical resources, on the other hand, resemble dictionaries that are helpful in understanding the meanings of words found in different text corpora. Combined, text corpora and lexical resources provide a comprehensive means to analyze and learn any language.

Text corpora contain real-world examples of how humans use language in different contexts. Lexical resources offer definitions and explanations of words contained in text corpora. Together, the two learners can see words in real usage and, at the same time, understand the meanings of words, their usage in actual sentences, and the relationships between them. It makes the learning of language and its analysis more effective. Figure 2-1 illustrates some data processing techniques that were used for a sentiment analysis tutorial. We define, discuss, and use each of these data processing techniques in the upcoming chapters. Stay tuned!

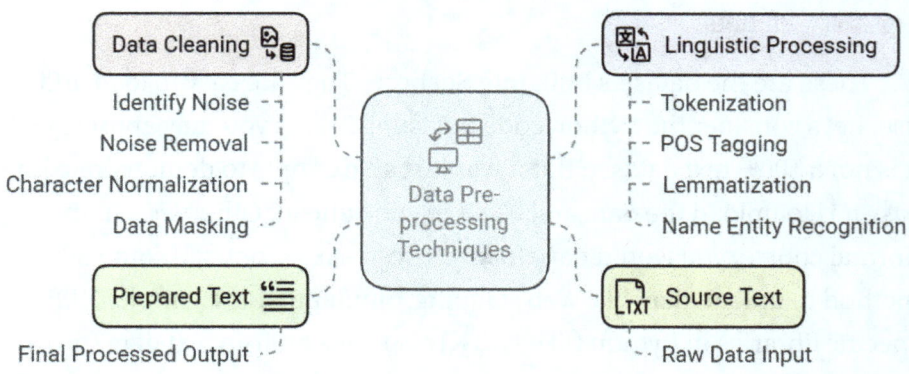

Figure 2-1. *Some common text preprocessing techniques in order*

In Figure 2-1, the source text is project-specific text corpora that can be used along with the required lexical resources if a project needs it. Next, let's discuss concepts and insert their real-time usage using Python code.

CHAPTER 2 COLLECTING AND EXTRACTING THE DATA FOR NLP PROJECTS

Our main objective in this chapter is to explain the sources of text data that you can use for NLP projects. Let's start with some familiar examples. If you are familiar with the Seaborn library in Python, probably you might have come across one or many of the following datasets (not necessarily text) available in the Seaborn library.

- Tips
- Iris
- Penguins
- Flights
- Diamonds
- Titanic
- Exercise
- MPG
- Planets

These are the datasets built into Seaborn. They are easy to load and use. Let's consider the Python code in Listing 2-1; as you may appreciate, it is not a 100% text dataset. But it was just an example to demonstrate how easy it is to upload the data and use it in code ahead. Otherwise, in the normal course, you would have loaded a .csv, a .xlxs, or used some other method to upload data like web scraping. Similar to this example, NLP-specific libraries in Python (like NLTK) also have built-in text data (text corpora) for easy use in NLP projects.

Listing 2-1. Easy uploading of internal data from Python Seaborn

```
import seaborn as sns
tips = sns.load_dataset("tips")
print(tips.head())
```

	total_bill	tip	sex	smoker	day	time	size
0	16.99	1.01	Female	No	Sun	Dinner	2
1	10.34	1.66	Male	No	Sun	Dinner	3
2	21.01	3.50	Male	No	Sun	Dinner	3
3	23.68	3.31	Male	No	Sun	Dinner	2

Parallel to the Seaborn example, let's now discuss some built-in text corpora from the NLTK Python library. You can readily use this text data in your analytics (NLP) projects.

NLTK is a leading NLP platform for building Python programs that work with natural languages, such as English, and others. NLTK has built-in, easy-to-use interfaces to more than 50 corpora and lexical resources. Some popular ones include Brown, Gutenberg, Reuters, Inaugural Address Corpus, and WordNet 3.0 (English). The following code (Listing 2-2) deals with accessing popular text corpora and lexical resources that are built into NLTK.

2.2.1 Accessing Some Popular NLTK Text Corpora

Listing 2-2. Accessing popular text corpora and lexical resources

```
# You can skip this code if you have already installed nltk
Python library
# You can run this. I am not including the output here for
space saving.
!pip install nltk
# Once you install nltk, you need to download the data.
# You can run this. I am not including the output here for
space saving.
import nltk
nltk.download('all')
```

2.2.2 Accessing Gutenberg Corpus

```
from nltk.corpus import gutenberg
print(gutenberg.fileids())
['austen-emma.txt', 'austen-persuasion.txt', 'austen-sense.
txt', 'bible-kjv.txt', 'blake-poems.txt', 'bryant-stories.txt',
'burgess-busterbrown.txt', 'carroll-alice.txt', 'chesterton-
ball.txt', 'chesterton-brown.txt', 'chesterton-thursday.txt',
'edgeworth-parents.txt', 'melville-moby_dick.txt', 'milton-
paradise.txt', ….continued]
```

The code output represents the collection of file IDs (file identifiers) that are available in the Gutenberg text corpus of NLTK.

Each file ID represents a text file that contains a literary work included in the Gutenberg text corpora.

```
# This code reads the text of a particular file
hamlet = gutenberg.words('austen-persuasion.txt')
print(hamlet[:50])
['[', 'Persuasion', 'by', 'Jane', 'Austen', '1818', ']',
'Chapter', '1', 'Sir', 'Walter', 'Elliot', ',', 'of',
'Kellynch', 'Hall', ',', 'in', 'Somersetshire', ',', 'was',
'a', 'man', 'who', ',', 'for', 'his', 'own', 'amusement',
',', 'never', 'took', 'up', 'any', 'book', 'but', 'the',
'Baronetage', ';', 'there', 'he', 'found', 'occupation', 'for',
'an', 'idle', 'hour', ',', 'and', 'consolation']
```

Let's access a few more such text corpora.

2.2.3 Accessing Reuters Corpus

```
from nltk.corpus import reuters
file_ids = reuters.fileids()[:10]
print(file_ids)
```

CHAPTER 2   COLLECTING AND EXTRACTING THE DATA FOR NLP PROJECTS

```
[....., 'test/14833', 'test/14839', 'test/14840', 'test/14841',
'test/14842', ....continued]
```

Categories in the Reuters corpus are not mutually exclusive. We can request the multiple topics covered by many documents. We can also request multiple documents included in several categories.

```
# The corpus procedures can accept a single or a multiple file
IDs at a time.
reuters.categories(['test/14828', 'test/14829'])
['crude', 'grain', 'nat-gas']
file_ids = reuters.fileids(['crude', 'grain'])[:10]
print(file_ids)
['test/14828', 'test/14829', 'test/14832', ...continued]
reuters.words('test/14843')[:5]
['SUMITOMO', 'BANK', 'AIMS', 'AT', 'QUICK']
reuters.words(categories=['crude', 'grain', 'nat-gas'])
['CHINA', 'DAILY', 'SAYS', 'VERMIN', 'EAT', '7', '-', ...]
```

2.2.4 Accessing Brown Corpus

For accessing Brown Corpus, the following code can be used.

```
from nltk.corpus import brown
categories = brown.categories()[:5]
print(categories)
[...., 'editorial', 'fiction', ...continued]
sentences = brown.sents(categories=['adventure',
'belles_lettres'])[:2]
for sentence in sentences:
    print(' '.join(sentence))
Northern liberals are the chief supporters of civil rights and
of integration .
They have also led the nation in the direction of a welfare state .
```

2.2.5 Accessing Gutenberg Corpus

Reuters Corpus has 25,000 free ebooks hosted on its website.

We can run nltk.corpus.gutenberg.fileids() to get the file IDs in the corpus.

```
import nltk
file_ids = gutenberg.fileids()[:3]
print(file_ids)
['austen-emma.txt', .... continued]
# We will take the text - "Emma by Jane Austen"
# We will be then named as "emma_text".
# After that we will find out the number of words in it.

emma_text_sample = nltk.corpus.gutenberg.words('austen-emma.txt')
len(emma_text_sample)
192427
```

2.2.6 Accessing Web and Chat Text

This corpus includes content from a Firefox discussion forum, the movie script for *Pirates of the Caribbean*, conversations overheard in New York, customized commercials, and reviews of wines.

For accessing Web and Chat Text, the following code can be used.

```
from nltk.corpus import webtext
for fileid in webtext.fileids():
    print(fileid, webtext.raw(fileid)[:30], '...')
firefox.txt Cookie Manager: "Don't allow s ...
grail.txt SCENE 1: [wind] [clop clop clo ...
...
...
pirates.txt PIRATES OF THE CARRIBEAN: DEAD ...
singles.txt 25 SEXY MALE, seeks attrac old ...
```

```
wine.txt Lovely delicate, fragrant Rhon ...
# A corpus of instant messaging chat sessions is also
available in nltk
# This corpus is organized into 15 files
```

```python
from nltk.corpus import nps_chat
chatroom = nps_chat.posts('10-19-20s_706posts.xml')
first_10_messages = chatroom[123][:10]
print(first_10_messages)
```

2.2.7 Accessing NLTK Lexical Resources

The following are some key points.

- NLTK contains multiple lexical resources.

- The most significant one is WordNet.

- WordNet is a huge lexical database of English.

- It groups words into multiple sets of synonyms.

    ```python
    from nltk.corpus import wordnet as wn
    synonyms = wn.synsets('book')
    print(synonyms)
    [Synset('book.n.01'), Synset('book.n.02'), ....
    continued.
    ```

- The output represents a list of synonym sets (called synsets)for the word "book".

- book represents the word for which the synonym set is defined.

- n represents the part of speech ("n" stands for noun in this specific example).

- 01 represents the sense number that differentiates the different meanings of the word.

```
# Get definitions and examples. Printing only the
first five.
for i, syn in enumerate(synonyms):
    if i >= 5:
        break
    print(syn.definition())
    print(syn.examples())
a written work or composition that has been published
(printed on pages bound together)
['I am reading a good book on economics']
physical objects consisting of a number of pages bound
together
['he used a large book as a doorstop']
a compilation of the known facts regarding something
or someone
["Al Smith used to say, `Let's look at the record'",
'his name is in all the record books']
a written version of a play or other dramatic
composition; used in preparing for a performance
[]
a record in which commercial accounts are recorded
['they got a subpoena to examine our books']
```

- The output gives the definitions and example sentences for each sense of the word *book*.

- It is retrieved from WordNet.

- It also illustrates the various contexts in which we can use the word.

- It also illustrates the various contexts in which we can use the word.

>>> *Code Snippet 2-2*

There are many methods available with NLTK that deal with basic corpus functionality. These methods can be found at nltk.corpus.reader and www.nltk.org/howto. You are strongly advised to visit these sites to get full details.

2.3 Extracting Data from Word Files

The use of Microsoft Word documents is widespread in professional, research, and academic environments. Word files often contain valuable data in textual formats. Word files can include business reports, literary articles, and research papers. This data can be used for multiple NLP projects, including information extraction, sentiment analysis, and text classification. Processing of Word files can yield rich, structured, and unstructured text content, which can be useful in building comprehensive NLP models. Such models can be used to automate and streamline business workflows to enable quick insights and data-driven decision-making.

The information in this format can be combined with data from databases, websites, and other text formats (or sources). Data integration from various sources can enhance the effectiveness and functionality of NLP applications.

Listing 2-3 demonstrates the creation of a sample file. We will use this file to extract data. We will use the python-docx library for both writing to and reading from the Word file.

2.3.1 Extracting Data from MS Word Files

Create a sample file.

```
# Install python-docx if you haven't done it already.
# I am not including the output of this code due to space
constraints.
!pip install python-docx
```

The following script creates a sample Word demo document containing a title, multiple paragraphs, bullet points, and a table.

```python
from docx import Document
from docx.shared import Pt

# Create a new blank Document.
doc = Document()

# Add a title
doc.add_heading('Sample Document For NLP Analysis', level=1)

# Add a couple of paragraphs.
doc.add_paragraph('This sample Word demo document contains multiple paragraphs.')
doc.add_paragraph('Here is another paragraph with more text.')

# Add few bullet points.
doc.add_paragraph('Sample bullet 1', style='List Bullet')
doc.add_paragraph('Sample bullet 2', style='List Bullet')
doc.add_paragraph('Sample bullet 3', style='List Bullet')

# Add a table for demo.
table = doc.add_table(rows=3, cols=2)
table.style = 'Table Grid'
```

CHAPTER 2   COLLECTING AND EXTRACTING THE DATA FOR NLP PROJECTS

```
# Add header row
hdr_cells = table.rows[0].cells
hdr_cells[0].text = 'Header Sample 1'
hdr_cells[1].text = 'Header Sample 2'

# Add a couple of data rows
row_cells = table.rows[1].cells
row_cells[0].text = 'Sample Row 1'
row_cells[1].text = 'Data Point 1'

row_cells = table.rows[2].cells
row_cells[0].text = 'Sample Row 1'
row_cells[1].text = 'Data Point 2'

# Save the newly created document.
doc.save('NLP_demo_sample_file.docx')
```

The following script loads this Word document.

Then, it extracts and prints various elements of the document, like the title, paragraphs, bullet points, and table data.

Listing 2-3. Extracting Data from MS Word Files

```
from docx import Document

# First load the Word document created above.
doc = Document('NLP_demo_sample_file.docx')

# Extract and print the title
# Assume it to be the opening paragraph.
print(f"Title: {doc.paragraphs[0].text}")

# Extract and print all paragraphs.
print("\nParagraphs:")
for para in doc.paragraphs:
    print(para.text)
```

```python
# Extract and print bullet points (assuming they are in a list)
print("\nBullet Points:")
for para in doc.paragraphs:
    if para.style.name == 'List Bullet':
        print(para.text)

# Extract and print table data
print("\nTable Data:")
for table in doc.tables:
    for row in table.rows:
        for cell in row.cells:
            print(cell.text, end=' | ')
        print()
```
Title: Sample Document For NLP Analysis

Paragraphs:
Sample Document For NLP Analysis
This sample Word demo document contains multiple paragraphs.
Here is another paragraph with more text.
Sample bullet 1
Sample bullet 2
Sample bullet 3

Bullet Points:
Sample bullet 1
Sample bullet 2
Sample bullet 3

Table Data:
Header Sample 1 | Header Sample 2 |
Sample Row 1 | Data Point 1 |
Sample Row 1 | Data Point 2 |

2.4 Extracting Data from HTML

A vast amount of structured and unstructured text data exists in the form of HTML documents. Websites, blogs, and online articles are often in HTML format. In such cases, HTML documents become one important primary source to extract real-world language usage and patterns. Parsing of HTML text can give access to a wealth of diverse textual data that is crucial for training NLP models.

This structured nature of HTML documents makes it relatively easier to extract appropriate information. The inherent metadata of HTML documents, such as tags and attributes (e.g., <title>, <h1>, and <p>), provide additional background to the text. HTML documents frequently include hyperlinks, which facilitates the exploration across the web. Listing 2-4 first creates a sample HTML file for the demo, and then we present well-commented code to extract various elements of the sample HTML document. Figure 2-2 depicts the data extraction from HTML documents.

Text Data Extraction from HTML

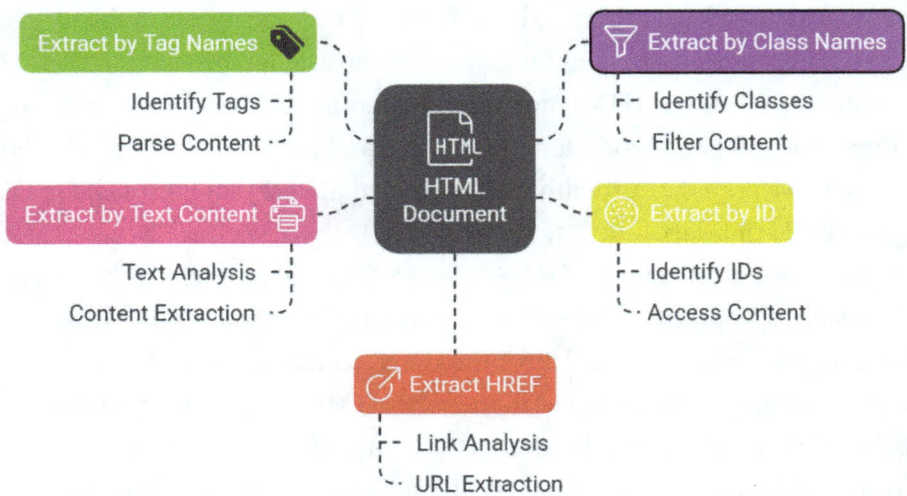

Figure 2-2. *The process of text extraction from HTML using the Beautiful Soup Python library*

2.4.1 Extracting Data from HTML Documents

First, let's create an HTML document for use later for extracting text out of it (see Listing 2-4). You should have some basic knowledge of HTML files to extract the best out of this section.

Listing 2-4. Extracting Data from HTML Documents using the Beautiful Soup library

```
# Import the BeautifulSoup class from the bs4 library
from bs4 import BeautifulSoup

# The following code reads the HTML file.
# Open 'Sample.html' in read mode.
with open('my_sample.html', 'r') as file:
```

```python
    html = file.read()  # Read the content of the file into the
    'html' variable

# Parse the HTML content using BeautifulSoup.
soup = BeautifulSoup(html, 'html.parser')

# Extract the text of the <title> element.
title = soup.title.string
# Extract the text of the <h1> element within the <header>.
main_heading = soup.header.h1.string
# Extract all href attributes from <a> elements within
the <nav>.
nav_links = [a['href'] for a in soup.nav.find_all('a')]
# Extract the text of the <p> element.
home_section = soup.find(id="home").p.string
# Extract the text of the <p> element within the section with
id="about".
about_section = soup.find(id="about").p.string
# Extract the text of the <p> element within the section with
id="contact".
contact_section = soup.find(id="contact").p.string
# Extract the href attribute of the <a> element.
contact_email = soup.find(id="contact").a['href']
# Extract the text of the <p> element within the <footer>.
footer_text = soup.footer.p.string

# Print the title.
print("Title:", title)
# Print the main heading.
print("Main Heading:", main_heading)
# Print navigation links.
print("Navigation Links:", nav_links)
# Print the text of the home section.
print("Home Section:", home_section)
```

CHAPTER 2 COLLECTING AND EXTRACTING THE DATA FOR NLP PROJECTS

```
# Print the contents of the about section.
print("About Section:", about_section)
# Print the contents of the contact section.
print("Contact Section:", contact_section)
# Print the contact email link.
print("Contact Email:", contact_email)
# Print the footer text.
print("Footer Text:", footer_text)
```
OUTPUT:
```
Title: Sample HTML File
Main Heading: Main Title
Navigation Links: ['#home', '#about', '#contact']
Home Section: Welcome home. Can we have a coffee?.
About Section: This section writes about an overview of my website.
Contact Section: None
Contact Email: mailto:my_name@my_email.com
Footer Text: © I am writing a book with my copyright on it.
```

You are encouraged to try out this code piece with any other HTML document.

```
# Create an HTML document as said.
html_content = '''
<!DOCTYPE html>
<html lang="en">
<head>
    <meta charset="UTF-8">
    <meta name="viewport" content="width=device-width, initial-scale=1.0">
    <title>Sample HTML File</title>
</head>
<body>
```

```html
    <header>
        <h1>Main Title</h1>
        <nav>
            <ul>
                <li><a href="#home">Home</a></li>
                <li><a href="#about">About</a></li>
                <li><a href="#contact">Contact</a></li>
            </ul>
        </nav>
    </header>
    <section id="home">
        <h2>Home Section</h2>
        <p>Welcome home. Can we have a coffee?.</p>
    </section>
    <section id="about">
        <h2>About Section</h2>
        <p>This section writes about an overview of my
        website.</p>
    </section>
    <section id="contact">
        <h2>Contact Section</h2>
    <p>You can contact me using my personal email id<a
    href="mailto:my_name@my_email.com">info@example.
    com</a>.</p>
    </section>
    <footer>
        <p>&copy; I am writing a book with my copyright
        on it.</p>
    </footer>
</body>
</html>
```
...

```
# Save the HTML content to a file
with open('my_sample.html', 'w') as file:
    file.write(html_content)
```

The Beautiful Soup library is used to extract text from this HTML file. It is a popular Python library to scrape data from the web.

2.5 Extracting Data from JSON

Like HTML text, JSON files also provide a structured data format for storing and exchanging data. JSON (JavaScript Object Notation) files are lightweight and easy to parse. It makes them almost a perfect option for handling large datasets that are typical in NLP projects. JSON files have a nested structure that supports complex data representations with relative ease. It allows for the effective organization of text, text annotations, and its metadata. All these properties of JSON files make them a preferred option when it comes to text preprocessing and analysis.

Let's create a sample JSON file and then present a well-documented code for data extraction.

2.5.1 Extracting Data from JSON Files

First, let's create JSON for use later for extracting text out of it (see Listing 2-5). You should have some basic knowledge of JSON files to extract the best out of this section.

Listing 2-5. Extracting Data from JSON Files

```
import json # import the required Python library to create the JASON file.

# Let's create a sample dictionary with some data.
data = {
```

```
    "title": "Sample JSON File",
    "description": "This is a sample JSON file containing
    various elements.",
    "sections": [
        {
            "heading": "Introduction to the World of JASON",
            "content": "This is your introduction to the
            content of a sample JSON file."
        },
        {
            "heading": "Details of JASON",
            "content": "Put your details about the JSON file
            content here."
        },
        {
            "heading": "Conclusion",
            "content": "Put your concluding remarks here."
        }
    ],
    "footer": "Put your footer text here."
}
# Save the sample JSON file,
with open('sample_jason.json', 'w') as file:
    json.dump(data, file, indent=4)
```

- Now let's extract its content.

```
import json # Import json.
```

```
# Read the JSON file.
with open('sample_jason.json', 'r') as file:
# Open the JASON file in read mode.
```

```
    data = json.load(file)  # Load the JSON file into the
    'data' variable.

# Extract various elements.
title = data["title"]  # Extract "title".
description = data["description"]  # Extract "description".
sections = data["sections"]  # Extract "sections".
footer = data["footer"]  # Extract "footer".

# Print the extracted text data.
print("Title:", title)  # Print the title.
print("Description:", description)  # Print the description.
for section in sections:  # Loop through the sections list.
    print(f"Section Heading: {section['heading']}")
# Print the section heading.
    print(f"Section Content: {section['content']}")
# Print the section content.
print("Footer:", footer)  # Print the footer text.
OUTPUT:
Title: Sample JSON File
Description: This is a sample JSON file containing various
elements.
Section Heading: Introduction to the World of JASON
Section Content: This is your introduction to the content of a
sample JSON file.
Section Heading: Details of JASON
Section Content: Put your details about the JSON file
content here.
Section Heading: Conclusion
Section Content: Put your concluding remarks here.
Footer: Put your footer text here.
```

2.6 Extracting Data from PDFs

PDF files are used for distributing and archiving documents quite often. They contain rich text data in a consistent format. PDF text frequently includes structured text, images, and metadata about the text that can be helpful for extracting and later analyzing text content. Almost every business domain, including finance, legal, and academia, is dependent on files in PDF format. This makes them an important data source for many NLP tasks. Figure 2-3 is a simplified schematic of text data extracted from PDF files.

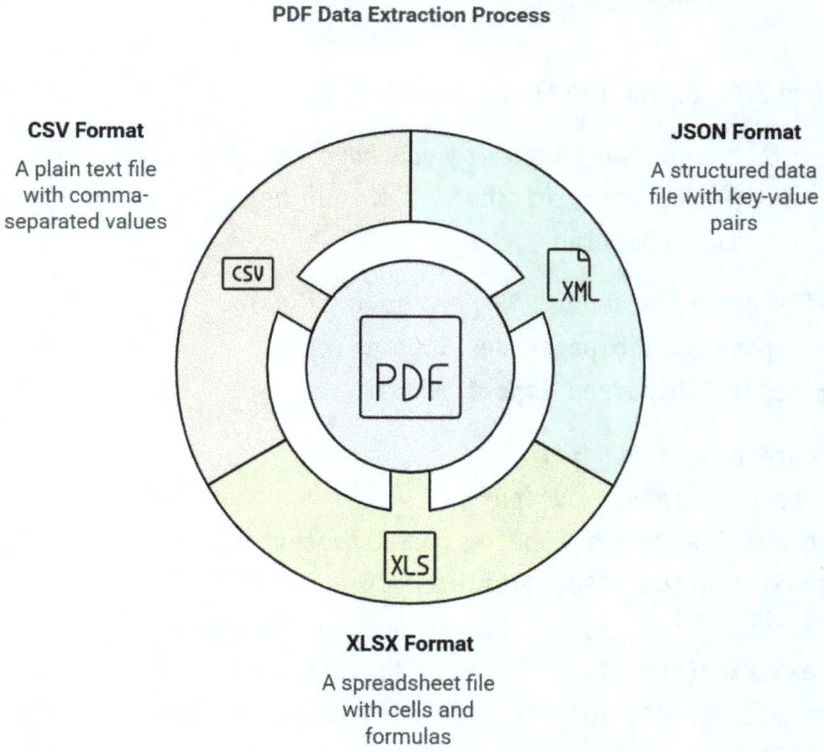

PDF Data Extraction Process

CSV Format
A plain text file with comma-separated values

JSON Format
A structured data file with key-value pairs

XLSX Format
A spreadsheet file with cells and formulas

Figure 2-3. *Text data extraction from PDF files*

2.6.1 Extracting Data from JSON Files

First, let's create a PDF file for use later for extracting text out of it (see Listing 2-6). You should have some basic knowledge of PDF files to get the best out of this section.

- We will first create a small PDF file for use later in extracting text out of it.
- We will use the reportlab library for creating the PDF.
- Later, we will use PyMuPDF (also called fitz) for extracting the data.

Listing 2-6. Extracting Data from PDFs

```
# Install reportlab module if you have not done it already.
# I am not including this code's output here for space reasons.
!pip install reportlab

# Let's create a small PDF demo file as said.
from reportlab.lib.pagesizes import letter
from reportlab.pdfgen import canvas

# Create a demo PDF file,
file_path = "sample_pdf.pdf"
c = canvas.Canvas(file_path, pagesize=letter)
c.drawString(100, 750, "Title: Demo PDF")
c.drawString(100, 735, "This is a demo PDF document.")
c.drawString(100, 720, "Section 1: Introduction")
c.drawString(100, 705, "Put your introductory text here.")
c.drawString(100, 690, "Section 2: Details")
c.drawString(100, 675, "Put the details of the PDF document.")
c.drawString(100, 660, "Section 3: Conclusion")
c.drawString(100, 645, "The conclusion text in this section.")
```

```python
c.save()
# To extract the PDF text install fitz if not done already.
# This code's output not included here.
!pip install PyPDF2
import PyPDF2

# Open the PDF file
file_path = "sample_pdf.pdf"
with open(file_path, 'rb') as file:
    reader = PyPDF2.PdfReader(file)
    number_of_pages = len(reader.pages)
    text = ""
    for page_number in range(number_of_pages):
        page = reader.pages[page_number]
        text += page.extract_text()
print("Extracted Text:")
print(text)
```

```
Extracted Text:
Title: Demo PDF
This is a demo PDF document.
Section 1: Introduction
Put your introductory text here.
Section 2: Details
Put the details of the PDF document.
Section 3: Conclusion
The conclusion text in this section.
```

2.7 Web Scraping

Web scraping enables the large collection of text data from the World Wide Web. This data is an important source of text data for training and evaluating NLP models. Researchers and practitioners can explore a variety of sources and build up diverse datasets that can improve an NLP model's ability to comprehend a variety of languages, contexts, and topics.

The World Wide Web is a source of up-to-date and real-time information. This dynamic data can be crucial in creating more accurate and relevant NLP applications by reflecting current language usage and emergent trends. Listing 2-7 is a simple example of how to scrap data from any web URL. Figure 2-4 depicts the entire process of generic web scraping.

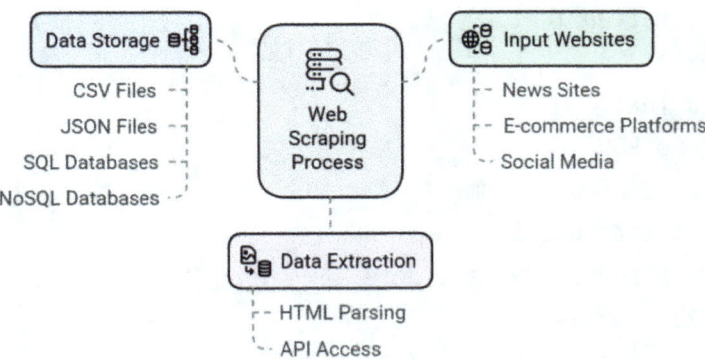

Figure 2-4. *Web scraping*

2.7.1 Web Scraping in NLP Tasks

Listing 2-7. Web Scraping in NLP Tasks

```python
import requests
from bs4 import BeautifulSoup

# URL of the webpage to scrape.
url = 'https://en.wikipedia.org/wiki/Blog'
# Inserting a random URL.

# Send a GET request to the URL
response = requests.get(url)

# Parse the HTML content of the scrapped page.
soup = BeautifulSoup(response.content, 'html.parser')

# Extract text data.
text_data = ''
for paragraph in soup.find_all('p'):
    text_data += paragraph.get_text() + '\n'

print('Scraped Text Data:')
print(text_data)
OUTPUT: Scraped Text Data (showing only the first paragraph):
```

A blog (a truncation of "weblog")[1] is an informational website consisting of discrete, often informal diary-style text entries (posts). Posts are typically displayed in reverse chronological order so that the most recent post appears first, at the top of the web page. In the 2000s, blogs were often the work of a single individual, occasionally of a small group, and often covered a single subject or topic. In the 2010s, "multi-author blogs" (MABs) emerged, featuring the writing of multiple authors and sometimes professionally edited. MABs from newspapers, other media outlets, universities, think tanks, advocacy groups, and similar institutions account for an increasing quantity of blog traffic. The rise of Twitter and other "microblogging" systems helps integrate MABs and single-author blogs into the news media. Blog can also be used as a verb, meaning to maintain or add content to a blog. Continued...

2.8 Recap of Key Concepts

NLP is an extremely influential tool used to analyze and comprehend large volumes of text data, which is abundantly available in all business domains. Text analysis using modern NLP techniques can reveal patterns, trends, and insights to make data-driven decisions across various fields, including business, legal, and social media.

Finding the right data, both in quality and quantity is one of the most challenging tasks in any NLP project. This chapter discussed various text data sources (corpora) and demonstrated how to extract meaningful data from them. This chapter is just a starting point for any NLP professional. Much more detailed and advanced techniques are available to source quality data for NLP projects. Internet and standard NLP textbooks are the best sources if you need more information on this topic. Web scraping is one of the most important processes because many large-scale AI (read GenAI) projects use it as their primary source of data for training their massive NLP models.

The chapter also discussed other techniques to source the quality data and techniques to extract text data from Word files, HTML documents, JSON files, and PDFs. We noticed that each format requires different method(s) to get the data into a usable format. Mastering these techniques is important for every NLP engineer as it's the starting point of any NLP project that they may indulge in.

2.9 Practice Exercises

1. Access the Gutenberg text Corpus that is distributed with NLTK. List all the book titles included in it. Practice text extraction and summarization with any two of its books.

2. Create a demo Microsoft Word document containing a few paragraphs of text data. Use a proper Python library to extract the text and print the data contained in it.

3. Repeat all the steps given in question 2 for an HTML file.

4. Practice data extraction from a real-time JSON file and a real web URL.

2.10 References

[1] https://www.oreilly.com/api/v2/epubs/9781492074076/files/assets/btap_0401.png. Last Accessed: July, 29 2024.

[2] Datacamp (2023). *NLTK Sentiment Analysis Tutorial for Beginners*. Available at: https://www.datacamp.com/tutorial/text-analytics-beginners-nltk

[3] nltk.org (2019). *Accessing Text Corpora and Lexical Resources*. Available at: https://www.nltk.org/book/ch02.html. Last Accessed: July 30, 2024.

[4] Tripathi, S. (2023). *How to parse HTML in Python with PyQuery or Beautiful Soup.* Available at: https://blog.apify.com/how-to-parse-html-in-python/. Last Accessed: July 30, 2024.

[5] Timalsina, A. (2024). *How to extract data from a PDF.* Available at: https://www.docsumo.com/blog/extract-data-from-pdf. Last Accessed: July 30, 2024.

[6] kinsta.com (2022). *What Is Web Scraping? How To Legally Extract Web Content.* Available at: https://kinsta.com/knowledgebase/what-is-web-scraping/. Last Accessed: July 30, 2024.

CHAPTER 3

NLP Data Preprocessing Tasks Involving Strings and Python Regular Expressions

3.1 Why You Should Read This Chapter

First, let's address our top-level objective. To keep it simple, we are looking at a new, hands-on skill in text analytics/natural language processing (NLP). This skill is currently in high demand in the global job market, as it's useful to all companies, regardless of their business domain. If we learn only the theoretical aspects or conduct a conceptual study, it does not provide us with the desired hands-on skills in NLP.

We need tools and techniques to analyze the large volumes of text data abundantly available in most establishments. We use NLP-specific and other Python libraries as a tool to get much-wanted actionable insights from volumes of unstructured text data. These Python libraries extensively use relevant statistical models for the analysis and evaluation of results. To gain hands-on skills, we must familiarize ourselves familiarized with these Python libraries and math algorithms. It all boils down to coding with Python, which you cannot master unless you get your hands dirty with it. So, get ready for some interesting action ahead in this chapter.

3.2 Python for Language Processing

Over the past few years, Python has emerged as a critical language for NLP due to its ease of use, wide range of NLP-specific libraries, and large user global community. Its syntax is simple to learn and use. All of this, combined with the easy readability of Python code, makes it a brilliant choice for both novices and experienced programmers. Python can efficiently handle large volumes of text data. A large number of instances are available where Python is used for NLP tasks such as text classification, sentiment analysis, language translation, and more. Python offers enough flexibility for integration with a variety of other machine learning models to further increase its utility in NLP tasks.

The NLP community uses several Python libraries that have become the backbone of NLP tasks due to their robust features and ease of use. NLTK (Natural Language Toolkit) is one such Python library that we use often. It provides easy-to-use tools for various NLP tasks such as tokenization, stemming, and tagging. spaCy is another Python library for NLP applications that is known for its speed and efficiency. spaCy finds its place in industrial-grade NLP applications. It supports deep learning workflows. TextBlob is another library used by beginners because it offers a simple API for common NLP tasks.

Other Python libraries, such as Genism, specialize in topic modeling and document similarity. Hugging Face is known for its state-of-the-art models, such as transformer models, which are extensively used in advanced NLP projects, including GenAI. Many of the terminology used in this paragraph may be new to you. We will cover everything either in this chapter or at appropriate places in the following chapters.

3.2.1 A Text Analytics Project Life Cycle (Generic NLP Pipeline)

In a typical machine learning (ML) project, there are steps such as data collection, preprocessing, feature extraction, model training, evaluation, and deployment. These steps, when done in series, one after the other, are often called an ML pipeline. In ML projects, data is frequently structured and consists of a combination of numerical and non-numeric (categorical) data. Such data typically requires preprocessing steps like normalization, scaling, and transformation. Once the data is ready, the ML model can be trained on this processed data. The model results are then evaluated using metrics like accuracy or F1 score. The final step is the deployment of the model into production for real-world predictions. An ML pipeline continuous monitoring and maintenance. Many steps in the pipeline are part of a cyclic process. Figure 3-1 depicts a general ML pipeline.

CHAPTER 3 NLP DATA PREPROCESSING TASKS INVOLVING STRINGS AND PYTHON REGULAR EXPRESSIONS

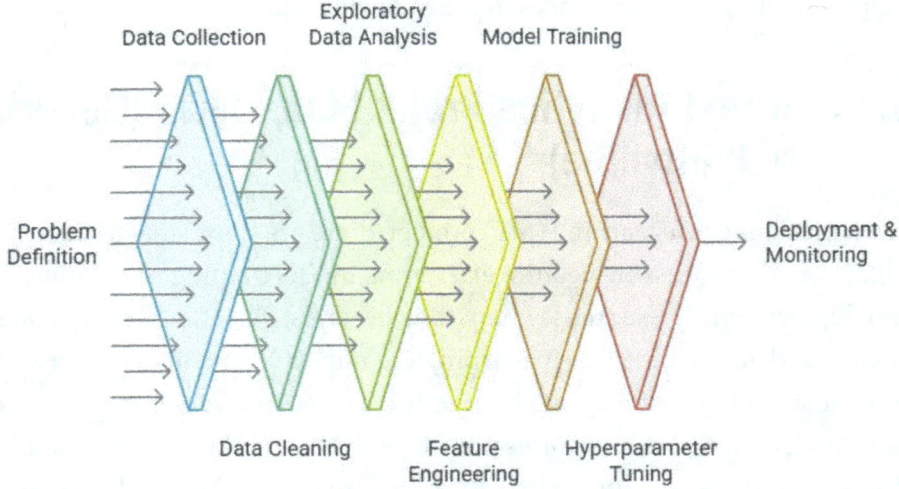

Figure 3-1. *A generic ML project pipeline (steps)*

An NLP project pipeline, on the other hand, deals with textual data. An NLP pipeline also involves data preprocessing steps, such as tokenization, stemming, lemmatization, and stop word removal. We will study each of these steps later at appropriate places. After these unique data preprocessing steps, text is converted into numerical representations, which is similar to ML pipelines. However, the techniques used in NLP pipelines are different. They are typically term frequency–inverse document frequency, word embeddings, or transformer models. You will learn about all of it later in this book. The processed textual data (converted to numeric representations) is then consumed by appropriate machine learning or deep learning models for training. A well-trained model can then accurately perform various NLP tasks, such as text classification, sentiment analysis, or named entity recognition.

By now, you might have noticed that many steps are similar in the training and deployment of ML and NLP models. The NLP pipeline typically includes several steps for processing text data. Operations and maintenance of trained models (in production) in both processes are similar and involve many processes done cyclically and repeatedly. We recommend that you refer to standard ML operations (MLOps) books for more details on the maintenance and operation of models in production use. Figure 3-2 depicts a generic NLP pipeline. You can compare Figures 3-2 and 3-1 to note similarities and differences in pipelines in both cases.

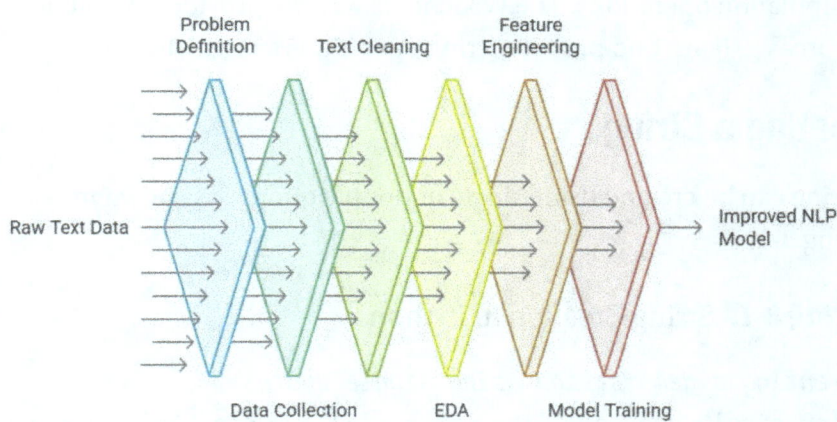

Figure 3-2. *A generic NLP project pipeline1 (steps)*

Many terms, especially in NLP pipelines, may sound unfamiliar to you at this moment. However, this chapter and the subsequent chapters focus on explaining these terms and providing hands-on skills to create and handle such pipelines.

Next, let's start with a basic piece of string handling and then proceed to relatively advanced topics of this chapter.

3.2.2 String Handling in Python

A String is a type of data structure in Python language. It characterizes a sequence of characters. A string is an immutable data type in Python. Once a string is created, it cannot be modified. Python strings are used extensively in most text applications. A string is used in storing and manipulating text data, which can be in the form of person names, local addresses, city names, and other data that can be denoted as text. Learning to handle string operations efficiently is basic to any NLP task. The following example uses Python code to demonstrate basic string manipulation operations. The W3Schools website provides excellent resources to learn and practice string operations using Python.

Creating a String

A string can be created using single or double quotes, as shown in Listing 3-1.

Listing 3-1. String Creation in Python

```
# Creating a demo string using single quotes.
string_single_quotes = 'This is a single quote string for demo'
print(string_single_quotes)
This is a single quote string for demo
# Creating a demo string using double quotes.
string_double_quotes = "This is a single quote string for demo"
print(string_double_quotes)
This is a double quote string for demo
```

You can create a multiline string (like a paragraph) using triple quotes (''' '''). You can try its Python code on your own. It's simple. If you are unable to crack it, refer to the W3Schools website. The following code demonstrations show how to create and manipulate text strings. These code snippets can be useful in data preprocessing for text analytics (NLP) steps.

CHAPTER 3 NLP DATA PREPROCESSING TASKS INVOLVING STRINGS AND PYTHON REGULAR EXPRESSIONS

Accessing Different Characters in a String

A string is represented as an array in Python. Figure 3-3 takes a string, "Python", and indexes its characters with numbers. The index position 0 represents the first character of the string, P. The last character, n, is represented by the index number 5 (Listing 3-2).

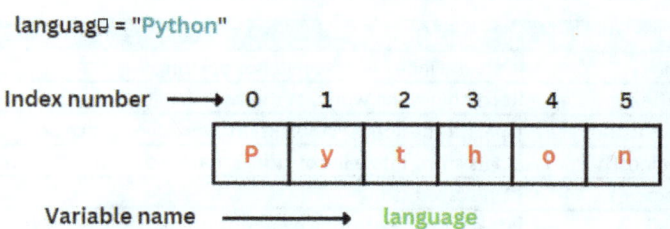

Figure 3-3. Representing characters of a string in Python

Listing 3-2. String Indexing in Python

```
# Accessing characters of a string by index bumbers
string_single_quotes = 'This is a single quote string for demo'
print(string_single_quotes)
This is a single quote string for demo
# Find the length of string 'This is a single quote string
for demo'.
len(string_single_quotes)
38
# Let's access the first "i" in this string.
string_single_quotes[2]
'i'
# Get the index number of the first "I" in string "'This is a
single quote string for demo".
string_single_quotes = 'This is a single quote string for demo'
index_of_i = string_single_quotes.index('i')
print("The index of the first 'i' is:", index_of_i)
The index of the first 'i' is: 2
```

CHAPTER 3   NLP DATA PREPROCESSING TASKS INVOLVING STRINGS AND PYTHON
            REGULAR EXPRESSIONS

More Basic String Operations

Table 3-1 gives a few sample string methods available in Python. For more details, please refer to W3Schools website. The code usage of additional methods is shown in Listing 3-3.

Table 3-1. *Selected Python String Methods*

Python string method	It's functionality
string.find(f)	Gets the index of first instance of string
string.join(text)	It combines the words of the text a single string
string.split(s)	Splits string into a Python list at every occurrence of "s"
string.splitlines()	Splits string into a list of strings, having only one at each line
string.titlecase()	Title cased string
string.strip()	Makes a copy of string by removing leading and trailing blanks
string.replace(k, t)	Replaces all instances of k with t inside the string

Listing 3-3. More basic string operations

```
# Slicing
for_slice_string = "String for slicing"
print(for_slice_string[2:5])
rin
# Slicing mid-way till end.
print(for_slice_string[11:])
slicing
# Slicing from beginning till mid-way.
print(for_slice_string[:6])
String
# Altering strings - converting in upper case.
demo_string = "this is a demo string in small letters"
print(demo_string.upper())
THIS IS A DEMO STRING IN SMALL LETTERS
# Converting in lower case
demo_string = "THIS IS A DEMO STRING IN capital LETTERS"
```

```
print(demo_string.lower())
this is a demo string in capital letters
```
 # Removing the extra blank spaces.
```
demo_string = " Remove blank spaces from this string "
print(demo_string.strip())
Remove blank spaces from this string
```
 # Replacing a substring within a string with an alternative substring.
```
demo_string = "Replace this with THAT"
print(demo_string.replace("this", "THAT"))
Replace THAT with THAT
```
 # Splitting a string at ','.
```
split_demo = "String, for, split, demo"
print(split_demo.split(",")) # Creates a Python list with individual words.
['String', ' for', ' split', ' demo']
```
 # Splitting a string at a specific character.
```
split_demo = "My house # is 7076 at #22nd street"
print(split_demo.split("#")) # Creates a Python list with individual words.
['My house ', ' is 7076 at ', '22nd street']
```

3.2.3 Introducing Regular Expressions for Text Processing

Regular expressions (RegEx) are essential tools in the preprocessing of text data. RegEx can extract specific patterns and clean text data in preparing data for training NLP models. This section introduces regular expressions and demonstrates how they can be used in data preprocessing.

CHAPTER 3   NLP DATA PREPROCESSING TASKS INVOLVING STRINGS AND PYTHON
           REGULAR EXPRESSIONS

A regular expression cut by a sequence of characters that formulates a search pattern. They can be used to determine if a text string matches the specified search pattern. Now, it is time to jump into code examples to understand RegEx better.

First, visit www.w3schools.com/python/python_regex.asp to get a list of RegEx metacharacters. Without these metacharacters, you will not be able to write or understand any RegEx code. Listing 3-4 is about only the Python RegEx methods.

Listing 3-4. Python Regular Expressions

```
Python RegEx examples.
# A list of all matches.

import re # the RegEx library.
# Below is a demo sentence.
demo_txt = "The referee had to match the players' skills to ensure a fair match."
# Find all occurrences of the word "match".
list = re.findall("match", demo_txt)
print(list)
['match', 'match']
# Split the string at each black space.

demo_txt = "The referee had to match the players' skills to ensure a fair match."
list = re.split("\s", demo_txt)
print(list)
['The', 'referee', 'had', 'to', 'match', 'the', "players'", 'skills', 'to', 'ensure', 'a', 'fair', 'match.']
# Let's now print the part of the string with a match.
```

```
demo_txt = "Last week I visited Spain"
# Visit W3school to find meaning of metacharacter "\bS\w+"
match = re.search(r"\bS\w+", demo_txt)
print(match.group())
Spain
```

The search functions. Search for the first white-space.

```
demo_txt = "I visited Europe and the US last year."
result_index = re.search("\s", demo_txt)
print("The location of first white-space character is at
index:", result_index.start())
The location of first white-space character is at index: 1
```
Replace each blank space character with "X".

```
demo_txt = "I visited Europe and the US last year."
result_string = re.sub("\s", "X", demo_txt)
print(result_string)
IXvisitedXEuropeXandXtheXUSXlastXyear.
```
Replace the first three blanks with "X"

```
demo_txt = "I visited Europe and the US last year. It was
freezing cold."
result_string = re.sub("\s", "X", demo_txt, 3)
print(result_string)
IXvisitedXEuropeXand the US last year. It was freezing cold.
```
Return a Matching Object.

```
demo_txt = "The rained in Mumbai last night. Mumbai is biggest
metro of India."
result_string = re.search("Mumbai", demo_txt)
print(result_string)
<re.Match object; span=(14, 20), match='Mumbai'>
```

Find the start-and end-position of the first match.

CHAPTER 3   NLP DATA PREPROCESSING TASKS INVOLVING STRINGS AND PYTHON
            REGULAR EXPRESSIONS

```python
import re

demo_txt = "The rained in entire last night. Spain is cold
these days."
result_string = re.search(r"\bS\w+", demo_txt)

if result_string:
    print(result_string.span())
else:
    print("No match found.")
(33, 38)
# Find the matching part of the string.

demo_txt = "The rained in entire Spain last night. Spain
is cold."
result_string= re.search(r"\bS\w+", demo_txt)\
print(result_string.group())
Spain
```

3.3 Real-life (NLP) Applications of RegEx

RegEx is handy when it comes to pattern matching and string manipulation in text analytics (or NLP) tasks. Python RegEx offers a quick method for matching, searching, and manipulating text based on predefined patterns. The following are selected real-life applications of Python RegEx. These code snippets can be used in data preprocessing and data cleaning steps to prepare data for the training of NLP models.

3.3.1 Validate Formatting of Phone Numbers

You can validate a list of phone numbers using RegEx to confirm they are in a precise format. For example, '234-567-9873'. Listing 3-5 is the Python code for this application.

Listing 3-5. Validating Phone Numbers with Python RegEx

```
import re

# Let's create a sample list of phone numbers.
# The valid phone number format is "123-456-789". Rest are
invalid.
phone_numbers_lst = [
    "567-789-7870",
    "986-666-3220",
    "1234567890",
    "135-42-7786",
    "xyz-abc-cdef"
]

# RegEx pattern for validating the list of phone numbers
# Refer W3school website on how to form a regular expression
pattern.
re_for_phone_pattern = r"^\d{3}-\d{3}-\d{4}$"

# Validate list phone numbers using this RegEx pattern.
for number in phone_numbers_lst:
    if re.match(re_for_phone_pattern, number):
        print(f"'{number}' is a valid phone number format.")
    else:
        print(f"'{number}' is invalid phone number format.")
'567-789-7870' is a valid phone number format.
'986-666-3220' is a valid phone number format.
```

'1234567890' is invalid phone number format.
'135-42-7786' is invalid phone number format.
'xyz-abc-cdef' is invalid phone number format.

3.3.2 Search and Replace

Any person who has used a spreadsheet or word processing application is familiar with the term "Search and Replace". Many other applications also make use of this functionality. This is handy when you need to make consistent changes across documents or codebases. Python RegEx module provides us with advanced search and replace functionality that enables accurate and flexible data manipulation (see Listing 3-6).

Listing 3-6. Python RegEx for Search and Replace

```
import re

# Let's first create a sample sentence.
demo_text = "The rain, rain, rain fell all day."

# We will replace the word "rain" with "water".

# REgEx pattern to search for "rain"
pattern = r"rain"

# Replace "rain" with "water"
replaced_demo_text = re.sub(pattern, "water", demo_text)

print("Original text:", demo_text)
print("Replaced text:", replaced_demo_text)
Original text: The rain, rain, rain fell all day.
Replaced text: The water, water, water fell all day.
```

3.3.3 Date Formatting

NLP professionals are often required to change the formatting of dates from one format to another. Due to the availability of various date formats, this operation can be challenging. The following code shows the possible simplification in this process by using Python's RegEx. In Listing 3-7, the application of regular expression is used to change the formats of dates from "MM-DD-YYYY" to "YYYY-MM-DD".

Listing 3-7. Python RegEx for Reformatting Dates

```
import re
```

```
# Sample date format in MM-DD-YYYY.
original_date_str = "10-08-2024"
```

```
# Python RegEx pattern to match tyhe original format -
MM-DD-YYYY.
original_pattern = r"(\d{2})-(\d{2})-(\d{4})"
```

```
# Let's reformat the sample date to another one - YYYY-MM-DD.
newly_formatted_date = re.sub(original_pattern, r"\3-\1-\2",
original_date_str)
```

```
print(newly_formatted_date)
2024-10-08
```

3.3.4 Word Counting

Word counting is finding the frequency of every word in a given text sentence or paragraph. Word count is useful for analyzing the content. The interest here is in finding the most frequently used words. Python RegEx module can be utilized for this purpose. Listing 3-8 demonstrates how to do it.

CHAPTER 3 NLP DATA PREPROCESSING TASKS INVOLVING STRINGS AND PYTHON
          REGULAR EXPRESSIONS

Listing 3-8. Python RegEx for Finding Word Frequencies

```
import re
from collections import Counter

# We will take the sample test as the first paragraph of this
section.
Sample_text_para = """
                   Word counting is finding the frequency
                   of every word in a given text sentence
                   or paragraph. Word count is useful for
                   analyzing of the content. The interest
                   here is in finding the most frequently used
                   words. Python RegEx module can be utilized
                   for this purpose. Below we will demonstrate
                   how to do it.
                   """

# RegEx pattern for matching words.
words = re.findall(r'\b\w+\b', Sample_text_para.lower())

# Find the frequency.
word_count = Counter(words)

print(word_count)
Counter({'the': 4, 'word': 3, 'is': 3, 'finding': 2, 'of': 2,
'in': 2, 'for': 2, 'counting': 1, 'frequency': 1, 'every':
1, 'a': 1, 'given': 1, 'text': 1, 'sentence': 1, 'or':
1, 'paragraph': 1, 'count': 1, 'useful': 1, 'analyzing':
1, 'content': 1, 'interest': 1, 'here': 1, 'most': 1,
'frequently': 1, 'used': 1, 'words': 1, 'python': 1, 'regex':
1, 'module': 1, 'can': 1, 'be': 1, 'utilized': 1, 'this': 1,
'purpose': 1, 'below': 1, 'we': 1, 'will': 1, 'demonstrate': 1,
'how': 1, 'to': 1, 'do': 1, 'it': 1})
```

3.3.5 Log File Analysis

Log file analysis is the extraction of specific information from large log files that are generated by servers and many other machines. Log files generally contain an exhaustive record of the activities performed by specific machines. We can use Python RegEx to parse log files and identify patterns, extract specific data, or filter out error records. RegEx can be used to match (and filter out) timestamps, error codes, or specific log entries. Extracting specific information from large log files can be helpful in statistical analysis. Log file analysis is often used to filter out ERROR records. Listing 3-9 demonstrates it with the Python RegEx module.

Listing 3-9. Python RegEx for Analysis of Log Files

```
import re

# We will first prepare an imaginary log file.
# And then filter out only the ERROR messages.
sample_log_data = """
2023-04-10 12:14:22 ERROR Could not connect to the server
2023-04-10 12:15:12 INFO User login timeout
2023-04-10 12:16:35 WARNING Out of memory
2023-04-10 12:17:34 ERROR Server module failure
"""

# RegEx pattern to match ERROR entries.
re_pattern_for_error = r"\d{4}-\d{2}-\d{2} \d{2}:\d{2}:\d{2} ERROR .*"

# Detect only the ERROR entries from the log file.
error_messages = re.findall(re_pattern_for_error, sample_log_data)

# Print each ERROR message one below the other.
```

```
for error in error_messages:
    print(error)
2023-04-10 12:14:22 ERROR Could not connect to the server
2023-04-10 12:17:34 ERROR Server module failure
```

3.3.6 String Cleaning

Text preprocessing tasks frequently need to clean the text strings. For example, a Tweet has a text string like Hello@John!, and our data cleaning task requires us to remove all special characters to make it usable for text analytics. We can deploy Python RegEx here, too. Using RegEx, you can remove special characters, white spaces, or URLs to clean up a text string for further analysis. Listing 3-10 is a demo.

Listing 3-10. Python RegEx for Cleaning Text Strings

```
import re

# Below is a sample text string, we found in a Tweet.
sample_text = "Hi!!! I am John. Welcome @my house. You can get more info @ www.sk.com"

# ReEx to remove special characters and URLs from the sample_text.
# Remove special characters
usable_text = re.sub(r"[!@#%&*:]", "", sample_text)  usable_text = re.sub(r"http\S+|www\.\S+", "", usable_text)  # Clean of URLs.

print(usable_text)
Hi I am John. Welcome my house. You can get more info
```

3.3.7 Tokenization

Tokenization is a central step in any NLP task. In this process, a larger text chunk is broken down into words or sentences to keep text analysis simple. Tokenization is principally a text preprocessing step that allows for preprocessing tasks like word frequency analysis, sentiment analysis, and language modeling. Python RegEx can be utilized for tokenization by first defining patterns to split the text chunks into smaller units of tokens based on mostly blank spaces, punctuation, or some other delimiters. Listing 3-11 is a code demonstration of a simpler method and a professional method.

Listing 3-11. Python RegEx for Tokenization of Larger Chunks of Text

- ***Simpler way to tokenize.***

```
import re # simpler way to tokenize.

# Below is a sample text chunk randomly cgosen for
tokenization.
sample_text = """ Tokenization is a central step in any
NLP task.
In this process, larger text chunk is broken down
into words
or sentences, to keep text analysis simple.
"""

# RegEx pattern to tokenize the text chunks into tokens
(words).,
tokens = re.findall(r'\b\w+\b', sample_text)

print(tokens)
```

```
['Tokenization', 'is', 'a', 'central', 'step', 'in', 'any',
'NLP', 'task', 'In', 'this', 'process', 'larger', 'text',
'chunk', 'is', 'broken', 'down', 'into', 'words', 'or',
'sentences', 'to', 'keep', 'text', 'analysis', 'simple']
```

- ***Below is more professional method to tokenize.***

```
import re # more professional method to tokenize.

def text_tokenizer(text_to_tokenize): # wtrite a function to toknize.
    """
    This function tokenizes any input text chunk into words
    by splitting on non-word characters.

    Args:
        text (str): The input string to tokenize.

    Returns:
        list: A list of tokens (words) extracted from the
        input text.
    """
    # RegEx pattern to split the text
    tokens = re.split(r'\W+', text_to_tokenize)

     Filter out any empty tokens that may result from the split.
    tokens = [token for token in tokens if token]

    return tokens
# Sample text for tokenization
sample_text = """ Tokenization is a central step in any NLP task.
In this process, larger text chunk is broken down into words
or sentences, to keep text analysis simple.
"""
```

```
# Tokenize the sample text.
tokens = text_tokenizer(sample_text)

# Output tokens are in the form of a list.
print(tokens)
['Tokenization', 'is', 'a', 'central', 'step', 'in', 'any',
'NLP', 'task', 'In', 'this', 'process', 'larger', 'text',
'chunk', 'is', 'broken', 'down', 'into', 'words', 'or',
'sentences', 'to', 'keep', 'text', 'analysis', 'simple']
```

3.3.8 Text Normalization

Formally speaking, *text normalization* refers to the conversion of larger text chunks into standard, consistent formats to minimize variations that can complicate text analysis tasks. Text normalization includes data cleaning tasks like converting text to lowercase, removing punctuation, expanding contractions, and replacing special characters. Text normalization ensures that different forms of a word or phrase are treated the same way. Text normalization cleans the text data into consistent forms to ensure better accuracy is attained when such cleaned data is used to train NLP models. Listing 3-12 is the text demo.

Listing 3-12. Python RegEx for Text Normalization

```
import re

def text_normalizer(text_to_normalize):
    """
    This function normalizes any input text by converting to
    lowercase,
    removing punctuation, and replacing special characters
    with spaces.

    Args:
```

```
    text (str): The input string to normalize.

Returns:
    str: The normalized text.
"""
# First convert the input text to lowercase.
text_to_normalize = text_to_normalize.lower()

# Replace special characters and punctuation with a
white space.
text_to_normalize = re.sub(r'[^\w\s]', ' ', text_to_
normalize)

# Remove any extra spaces.
text_to_normalize = re.sub(r'\s+', ' ', text_to_
normalize).strip()

    return text_to_normalize
# Sample input text.
sample_text = "Hi!!! This is Stella here. I'll come this
Evening with you to market!!!!!!"

# Normalize the text
normalized_text = text_normalizer(sample_text)

# Print the output.
print(normalized_text)
hi this is stella here i ll come this evening with you
to market
```

This section revealed how RegEx is useful. The RegEx code demos can help you see countless applications of RegEx in real-life situations. However, in every application of RegEx that we demonstrated so far, we did not discuss how a specific RegEx pattern can be cut for any explicit task. We suggest that you refer to the W3school website. There are also many good articles available on the Internet that can teach you.

CHAPTER 3   NLP DATA PREPROCESSING TASKS INVOLVING STRINGS AND PYTHON REGULAR EXPRESSIONS

3.4 Chapter Recap

This chapter began by understanding the significance of Python as a versatile tool for NLP tasks. Python is widely adopted in the industry due to its powerful libraries (whether for NLP or otherwise) and strong community support worldwide. We then investigated the generic NLP pipeline and examined various stages of a generic text analytics project. Then, we turned our focus on string handling in Python. While studying various string handling methods, you learned how to effectively manipulate and analyze text strings. We also studied regular expressions, focusing on their applications in data preprocessing tasks. From data cleaning to text extraction, the Python RegEx module has proved to be an irreplaceable asset in the text processing toolkit. This solid foundation in Python for NLP tasks will serve as a stepping stone for more advanced topics in language processing and text analytics. Throughout the following chapters, our goal is to provide the right balance of concepts and hands-on skills. Stay tuned!

3.5 Exercises

1. This chapter focuses on hands-on skills, primarily with regular expressions (RegEx). Write the Python RegEx programs for the following NLP data preprocessing tasks.

    - Email address validation
    - URL parsing
    - Extracting dates from text
    - HTML tag removal
    - Password strength checking

- Validating credit card numbers
- Extracting sentences from paragraphs
- Detecting duplicate words

2. From exercise question 1 and the examples given in this chapter, you may have discovered that sometimes, forming the right RegEx pattern for specific tasks can be tedious. In such cases, it's better to find alternate ways of doing the tasks. Search the Internet and find out if you can get alternate but easier ways of doing the tasks discussed in this chapter. Focus on real-life applications of the Python RegEx module.

3.6 References

[1] NTNU (2020). *NLP Pipeline #*. Available at: https://alvinntnu.github.io/python-notes/nlp/nlp-pipeline.html. Last Accessed: August 7, 2024.

[2] Oracle Cloud (2023). *Pipelines*. Available at: https://docs.oracle.com/en-us/iaas/data-science/using/pipelines-about.htm. Last Accessed: August 7, 2024.

[3] W3Schools (2024). Available at: https://www.w3schools.com/python/python_strings.asp. Last Accessed: August 7, 2024.

[4] Scientech Easy (2023). Available at: https://www.scientecheasy.com/2023/01/strings-in-python.html/. Last Accessed: August 7, 2024.

[5] W3Schools (2024). *Python RegEx*. Available at: https://www.w3schools.com/python/python_regex.asp. Last Accessed: August 7, 2024.

CHAPTER 4

NLP Data Preprocessing Tasks with NLTK

4.1 Why You Should Read This Chapter

Natural Language Toolkit (NLTK) is one of the most popular and useful natural language processing (NLP) platforms, useful for developing Python programs that work with natural languages (such as English, German, and others). NLTK offers user-friendly interfaces to more than 50 text corpora and other lexical resources. NLTK resources include WordNet, along with a collection of text processing libraries that can be useful for text classification, tokenization, stemming, tagging, parsing, and semantic reasoning. Additionally, there is an active discussion forum that can be very supportive and offer solutions to commonly occurring technical problems.

NLTK is free and open source. It is actively used by linguists, industry engineers, and researchers, university students, educators alike. Widespread online forums, tutorials, and example codes are available for supporting NLTK. It provides full support for the English language, but support for other languages like Spanish or French is not extensive.

Professionals often use NLTK with other useful Python libraries like scikit-learn (open source) and TensorFlow (open source from Google). It allows for even more sophisticated language modeling using deep learning and other machine learning algorithms. The chapter presents some applications of NLTK (using Python code) in data preprocessing and other NLP tasks. In each case, we keep a balance between the explanations and the required Python code.

The chapter also features a few simple NLP tasks that we can perform using NLTK.

4.2 NLP Data Preprocessing Tasks

NLP data preprocessing tasks include the steps necessary in preparing raw text data for NLP model training. Preprocessing tasks include data cleaning, structuring, and transforming the raw text to ensure consistency and maintain the required data quality. This step makes raw text data ready for model training and other more intricate NLP operations. Common preprocessing tasks include tokenization, removal of stop words, stemming, and lemmatization. Tokenization breaks down larger text chunks into individual words or sentences. Another operation, the removal of stop words, filters out common but irrelevant words. Stemming and lemmatization reduce words to their base or root forms. It ensures that multiple variations of the same word are treated uniformly during text analysis.

Let's start with tokenization using NLTK. As usual, we include a working code demo for all the processes.

4.2.1 Tokenization Using NLTK

Now that you have learned a bit about tokenization with RegEx, you will notice that it is much easier to tokenize with NLTK.

CHAPTER 4   NLP DATA PREPROCESSING TASKS WITH NLTK

Tokenization is a vital process in NLP data preprocessing. It involves breaking down larger text data into minor units of words or sentences. This operation further enables the identification of meaningful elements within the larger text chunks.

Tokenization can be at either the individual word level (word tokenization) or at the sentence level (sentence tokenization). Tokenization operations help in the proper structuring of unstructured raw text data to make it easier to process downstream NLP tasks. Listing 4-1 shows how to perform tokenization with just a few steps of code in NLTK.

Listing 4-1. Tokenization Using NLTK

```
# Install nltk if you have not done it already.
!pip install nltk # The code output not included here.
import nltk
from nltk.tokenize import word_tokenize, sent_tokenize

# Create sample text.
sample_text = """
We had already discussed about tokenization in RegEx sections.
Tokenization is a vital process in NLP data pre-processing.
It involves breaking down larger text data into minor units of
words or sentences.
This operation further enables the identification of meaningful
elements within the larger text chunks.
"""

# Perform word level tokenization first.
words = word_tokenize(sample_text)
print("Word Tokens:", words)

# Then perform sentence level tokenization.
sentences = sent_tokenize(sample_text)
```

```
print("\nSentence Tokens:", sentences)
OUTPUT:
Word Tokens: ['We', 'had', 'already', 'discussed', 'about',
'tokenization', 'in', 'RegEx', 'sections', '.', 'Tokenization',
'is', 'a', 'vital', 'process', 'in', 'NLP', 'data', 'pre-
processing', '.', 'It', 'involves', 'breaking', 'down',
'larger', 'text', 'data', 'into', 'minor', 'units', 'of',
'words', 'or', 'sentences', '.', 'This', 'operation',
'further', 'enables', 'the', 'identification', 'of',
'meaningful', 'elements', 'within', 'the', 'larger', 'text',
'chunks', '.']

Sentence Tokens: ['\nWe had already discussed about
tokenization in RegEx sections.', 'Tokenization is a vital
process in NLP data pre-processing.', 'It involves breaking
down larger text data into minor units of words or sentences.',
'This operation further enables the identification of
meaningful elements within the larger text chunks.']
```

4.2.2 Stop Word Removal

Stop word removal involves filtering out commonly used words (in sentences) that are often irrelevant to NLP tasks. These words are known as *stop words*. Examples of English stop words include "is," "the," "and," which typically do not contribute much to the meaning of sentences. By removing these words, the text analysis process can focus on more meaningful words. Removing stop words improves the efficiency and effectiveness of text analytics tasks such as text classification and sentiment analysis.

NLTK includes a list of English stop words. You can get them at https://gist.github.com/sebleier/554280. Stop words can vary depending on the language and context. The list of stop words provided

CHAPTER 4 NLP DATA PREPROCESSING TASKS WITH NLTK

in earlier sentences is a standard set of stop words for English. Stop word removal provides a clear advantage by reducing the dimensionality of text data and allows the analyst to focus on the most relevant words for analysis. Listing 4-2 is a code demo of stop word removal using NLTK.

Listing 4-2. Stop Word Removal Using NLTK

```
# Install nltk if you have not done it already.
!pip install nltk # The code output not included here.
# Important required libaries.
import nltk
from nltk.corpus import stopwords
from nltk.tokenize import word_tokenize

# Ensure required resources.
nltk.download('stopwords')
nltk.download('punkt')

def stopwords_remover(text, language='english'):
    """
    This functions filters out stop words from the input text.

    Args:
        text (str): The Sample text to remove stop words.
        language (str): The target language.

    Returns:
        list: A list of words with stop words removed.
    """
    # Tokenize the text.
    words = word_tokenize(text)

    # Retrieve the list of stop words for English.
    stop_words = set(stopwords.words(language))
```

CHAPTER 4　NLP DATA PREPROCESSING TASKS WITH NLTK

```python
    # Remove stop words.
    ouput_text = [word for word in words if word.lower() not in stop_words]

    return ouput_text
```
```
[nltk_data] Downloading package stopwords to C:\Users\Shailendra
[nltk_data]     Kadre\AppData\Roaming\nltk_data...
[nltk_data]   Package stopwords is already up-to-date!
[nltk_data] Downloading package punkt to C:\Users\Shailendra
[nltk_data]     Kadre\AppData\Roaming\nltk_data...
[nltk_data]   Package punkt is already up-to-date!
```
```python
# We will keep the sample text same as the tokenization example.
# This is to show the difference between tokenization and stop words removal.
sample_text = """
We had already discussed about tokenization in RegEx sections.
Tokenization is a vital process in NLP data pre-processing.
It involves breaking down larger text data into minor units of words or sentences.
This operation further enables the identification of meaningful elements within the larger text chunks.
"""

# Filter out stop words.
filtered_words = stopwords_remover(sample_text)

# Output the results.
print("Filtered Words:", filtered_words)
```
OUTPUT:

```
Filtered Words: ['already', 'discussed', 'tokenization',
'RegEx', 'sections', '.', 'Tokenization', 'vital', 'process',
'NLP', 'data', 'pre-processing', '.', 'involves', 'breaking',
'larger', 'text', 'data', 'minor', 'units', 'words',
'sentences', '.', 'operation', 'enables', 'identification',
'meaningful', 'elements', 'within', 'larger', 'text',
'chunks', '.']
```

> **Note** Compare the code output for tokenization and stop word removal. Do you find any difference? Discuss it in the class.

4.2.3 Stemming Using NLTK

Stemming might be a new word for most of you if you have not studied NLP before. This is a data preprocessing technique, and it comes under the broader category of normalization. This operation reduces words (as they appear in sentences of text corpora) to their base or root form. The transformed words are known as the stem. Stemming involves stripping suffixes like "ing," "ed," and "ly" to transform words into their base form. Stemming ensures that variations of a word are treated as the same during text analysis. Stemming also reduces the dimensionality of the data, like the operation of stop word removal (see Figure 4-1).

CHAPTER 4 NLP DATA PREPROCESSING TASKS WITH NLTK

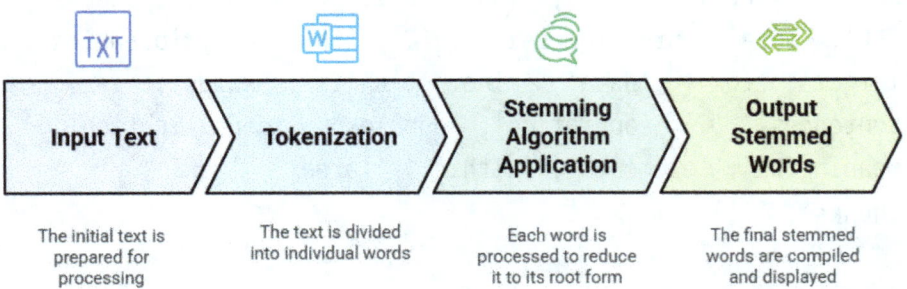

Figure 4-1. Stemming process with NLTK

Stemming can sometimes produce non-dictionary words, as demonstrated in the following examples.

- "Caring" can be stemmed to "car" in place of its root word, "care".

- "Happiness" can be stemmed to "happi" in place of its root word, "happy".

In spite of this approximation, stemming still remains a powerful tool for cultivating the efficiency and usefulness of text processing. You can access more examples directly from IBM's website at www.ibm.com/think/topics/stemming.

Listing 4-3 demonstrates stemming operation with NLTK. This function can be directly utilized in your NLP projects as one of the primary data preprocessing steps.

Listing 4-3. Stemming Using NLTK

```
# Install nltk if you have not done it already.
!pip install nltk # The code output not included here.
# Import required libraries.
import nltk
```

```python
from nltk.stem import PorterStemmer
from nltk.tokenize import word_tokenize

# Import other necessary resources.
nltk.download('punkt')

def stemmer(text):
    """
    Stems the input words using the Porter stemmer.

    Args:
        text (str): The original text to be stemmed.

    Returns:
        list: A list of stemmed words.
    """
    # Initialize the Porter Stemmer.
    stemmer = PorterStemmer()

    # Tokenize the input text.
    words = word_tokenize(text)

    # Apply stemmer to every word in the input text.
    stemmed_words = [stemmer.stem(word) for word in words]

    return stemmed_words
[nltk_data] Downloading package punkt to C:\Users\Shailendra
[nltk_data]     Kadre\AppData\Roaming\nltk_data...
[nltk_data]   Package punkt is already up-to-date!
# Create a random sample sentence for stemming.
sample_text = "NLP is language Processing that includes
analyzing, processing, and producing sentences."

# Apply the stemmer function, we just created.
stemmed_words = stemmer(sample_text)
```

```
# Print the resulting stemmed words.
print("Stemmed Words:", stemmed_words)
OUTPUT:
Stemmed Words: ['nlp', 'is', 'languag', 'process', 'that',
'includ', 'analyz', ',', 'process', ',', 'and', 'produc',
'sentenc', '.']
```

> **Note** Compare the output of the stemmed word with the original input sample text. Do you find a few non-dictionary words in the output? Discuss it in the class. Search the Internet to find if there are any better ways to reduce input words into their base forms.

4.2.4 Lemmatization

Lemmatization is an NLP data preprocessing technique that reduces input words to their base or root form. The base form of lemmatization is known as a lemma. Lemmatization is similar to stemming in a way. Lemmatization is crucial in NLP data preprocessing tasks like text normalization, where preserving the semantic meaning of original input words is critical (see Figure 4-2).

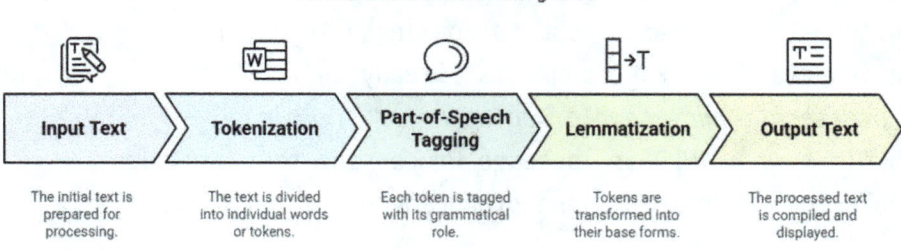

Figure 4-2. *Lemmatization using NLTK*

CHAPTER 4 NLP DATA PREPROCESSING TASKS WITH NLTK

While stemming simply truncates input words to remove only suffixes, lemmatization also considers the context and morphological analysis of words. Let's try to understand it better with an example. The input words "running," "ran," and "runs" can all be reduced to their lemma, "run." Consider the following paired text bullet sets.

- Original: running, stemmed: run
- Original: ran, stemmed: ran
- Original: runs, stemmed: run

and

- Original: running, Lemmatized: run
- Original: ran, Lemmatized: run
- Original: runs, Lemmatized: run

Let's repeat it with different examples.

- Original: easily, Stemmed: easili
- Original: fairly, Stemmed: fairli
- Original: happiness, Stemmed: happi

and

- Original: easily, Lemmatized: easily
- Original: fairly, Lemmatized: fairly
- Original: happiness, Lemmatized: happiness

With this example, you now have a clear picture of the difference between "stemming" and "lemmatization".

Now it's time to jump straight to our code demo on lemmatization works. The following code uses POS (part of speech), which means the role any word plays in a sentence. The role of a word (in a given sentence) may be a noun, verb, adjective, or adverb.

Considering POS during lemmatization is critical as it can have different base forms depending on their role/POS in a sentence. Let's take an example; "running" as a verb, becomes "run," but as a noun, it stays "running." In another example, if we use "better" as an adjective, it stays "better," but when used as a verb, it becomes "improve." Using POS, a lemmatizer can choose the correct base form. It preserves the meaning of the original input word. Listing 4-4 is a code demo for this operation. You can utilize this code directly in your NLP data preprocessing tasks.

Listing 4-4. Lemmatization Using NLTK

```
# Install nltk if you have not done it already.
!pip install nltk # The code output not included here.
import nltk # Import necessary libraries.
from nltk.stem import WordNetLemmatizer
from nltk.corpus import wordnet

# Import other required resources.
nltk.download('wordnet')
nltk.download('omw-1.4')
nltk.download('averaged_perceptron_tagger')

# Initialize the WordNet Lemmatizer.
lemmatizer = WordNetLemmatizer()

# Let's write a function to get part of speech for accurate
lemmatization
def get_wordnet_pos(word):
    tag = nltk.pos_tag([word])[0][1][0].upper()
    tag_dict = {"J": wordnet.ADJ,
                "N": wordnet.NOUN,
                "V": wordnet.VERB,
                "R": wordnet.ADV}
    return tag_dict.get(tag, wordnet.NOUN)
```

```
# The output of this code is not included here being
irrelavent.
# Sample imput words.
words = ["easily", "fairly", "happiness"]

# Lemmatization with considering part of speech (POS).
lemmatized_words = [lemmatizer.lemmatize(word, get_wordnet_
pos(word)) for word in words]

# Print the output.
for word, lemma in zip(words, lemmatized_words):
    print(f"Original: {word}, Lemmatized: {lemma}")
OUTPUT:
Original: easily, Lemmatized: easily
Original: fairly, Lemmatized: fairly
Original: happiness, Lemmatized: happiness
```

4.2.5 Sentence Segmentation

Sentence segmentation involves breaking an input text corpus into its component sentences. This process is an NLP data preprocessing task. It is useful where understanding the structure of the text is necessary. After the sentence segmentation process, the input text can be processed in the form of individual sentences. This operation helps in effective analysis (understanding meaning and context) of text corpora at the sentence level.

In sentence segmentation, sentence boundaries are determined by punctuation marks such as periods, exclamation points, and questions. However, special cases like abbreviations, decimal numbers, and names with punctuation marks do not indicate the end of a sentence.

The Python NLTK library can deal with sentence segmentation using its sent_tokenize function. The sent_tokenize function uses pre-trained NLP models to split text into sentences. It can deal with complicated sentence structures and edge cases. Listing 4-5 is a code demo for sentence segmentation. You can utilize this code directly in your NLP data preprocessing tasks.

Listing 4-5. Sentence Segmentation Using NLTK

```
# Install nltk if you have not done it already.
!pip install nltk # The code output not included here.
import nltk
from nltk.tokenize import sent_tokenize

nltk.download('punkt')

def sentences_segmenter(text):
    """
    Segments a given input text corpora into sentences.

    Parameters:
    text (str): The input text corpora to be segmented.

    Returns:
    list: A list of sentences extracted from the input text.
    """
    return sent_tokenize(text)

# The output of this code is not included here being irrelavent.
# Example text
text = """
```

Sentence segmentation is involves breaking an input text corpus into its component sentences.
This process is NLP data preprocessing task. It us useful where understanding the structure of the text is necessary. After sentence segmentation process, the input text can

CHAPTER 4   NLP DATA PREPROCESSING TASKS WITH NLTK

be processed in the form of individual sentences. This operation helps in effective analysis (understanding meaning and context) of text corpora at the sentence level.
"""

 # Segment the text corpus.
sentences = sentences_segmenter(text)

 # Print output.
for i, sentence **in** enumerate(sentences, 1):
    print(f"Sentence {i}: {sentence}")
OUTPUT:
Sentence 1: Sentence segmentation is involves breaking an input text corpus into its component sentences.
Sentence 2: This process is NLP data pre-processing task.
Sentence 3: It us useful where understanding the structure of the text is necessary.
Sentence 4: After sentence segmentation process, the input text can be processed in the form of individual sentences.
Sentence 5: This operation helps in effective analysis (understanding meaning and context) of text corpora at the sentence level.

4.2.6 Word Frequency Distribution

Word frequency distribution (WFD) comprises counting how often each word appears in an input text corpus. WFD is useful in understanding the importance of different words. It also helps in finding trends within the text. By exploring WFD, analysts can gain insights into the content and themes of the text.

In practical applications, WFDs are used in various NLP preprocessing tasks like keyword extraction, topic modeling, and sentiment analysis. We talk about these newly introduced terms later in this book. By knowing WFDs in an input text corpus, we can highlight important concepts or terms contained in that input text document. Comparing word frequencies across different input texts reveals the differences in their vocabulary and style (see Figure 4-3).

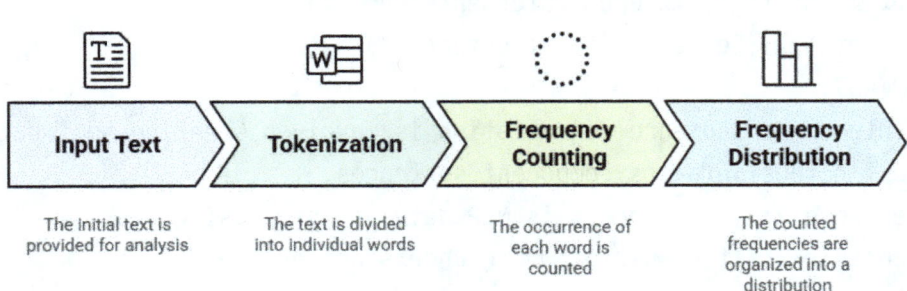

Figure 4-3. *WFD using nltk*

We can use the NLTK FreqDist class to create a frequency distribution for a list of words and analyze the most commonly occurring words in a given text corpus. This functionality is predominantly needed for processing large text corpora and extracting meaningful patterns contained in them. Listing 4-6 is a code demo for WFD. You can utilize this code directly in your NLP data preprocessing tasks.

Listing 4-6. WFD Using NLTK

```
# Install nltk if you have not done it already.
!pip install nltk # The code output not included here.
import nltk
from nltk.probability import FreqDist
from nltk.tokenize import word_tokenize
```

```python
nltk.download('punkt')

def word_frequency_finder(text):
    """
    Computes the WFD.

    Parameters:
    text (str): The input text.

    Returns:
    FreqDist: An WFD for given input words.
    """
    # Tokenize the text.
    words = word_tokenize(text)
    # Create WFD.
    freq_dist = FreqDist(words)
    return freq_dist
# The output of this code not included as it is irrelavent.
# Create sample text.
sample_text = """

In practical applications, WFDs are used in various NLP
pre-processing tasks
like keyword extraction, topic modeling, and sentiment
analysis. We will talk
about these newly introduces terms later in this book.

"""

# Compute WFD.
frequency_distribution = word_frequency_finder(sample_text)
```

```
# Print the output.
for word, frequency in frequency_distribution.items():
    print(f"Word: {word}, Frequency: {frequency}")
OUTPUT:
Word: ,, Frequency: 3:
Word: in, Frequency: 2................
```

4.2.7 Antonym and Synonym Detection Using NLTK

NLTK provides tools for these operations through WordNet, which is a lexical database to organize words into synsets (groups of synonyms). We leverage WordNet to find out antonyms as well. Listing 4-7 demonstrates how easily implement NLTK functions to detect both synonyms and antonyms.

Listing 4-7. Synonym Detection Using NLTK

```
# Install nltk if you have not done it already.
!pip install nltk # The code output not included here.
import nltk
from nltk.corpus import wordnet

# Ensure that the WordNet data is downloaded
nltk.download('wordnet')

def get_synonyms_and_antonyms(word):
    synonyms = set()
    antonyms = set()

    # Get synsets for the given word
    for syn in wordnet.synsets(word):
        # Add synonyms
```

CHAPTER 4   NLP DATA PREPROCESSING TASKS WITH NLTK

```
    for lemma in syn.lemmas():
        synonyms.add(lemma.name())
        # Add antonyms
        if lemma.antonyms():
            antonyms.add(lemma.antonyms()[0].name())

    return synonyms, antonyms
```

```
[nltk_data] Downloading package wordnet to C:\Users\Shailendra
[nltk_data]     Kadre\AppData\Roaming\nltk_data...
[nltk_data]   Package wordnet is already up-to-date!
# Sample word,
word = "Fast"
```

```
synonyms, antonyms = get_synonyms_and_antonyms(word)
```

```
print(f"Synonyms of {word}: {synonyms}")
print(f"Antonyms of {word}: {antonyms}")
Synonyms of Fast: {'degraded', 'riotous', 'dissipated', 'fast',
'firm', 'immobile', 'debauched', 'quick', 'truehearted',
'profligate', 'loyal', 'libertine', 'degenerate', 'fasting',
'dissolute', 'flying', 'tight'}
Antonyms of Fast: {'slow'}
```

4.2.8 Word Similarity Calculation

Word similarity calculation measures with how two words are similar in meaning. It uses definitions or usage of a word in text to compare their similarities. Let's try to understand it better with a simple example: "car" and "automobile" are similar as they both refer to the same type of vehicle; "cat" and "kitten" are similar as they refer to types of animals. Similarly, "computer" and "laptop" are related as they represent similar computing devices. Comparing the meanings, one can find out which words are similar (or even synonymous).

Word similarity can be measured in multiple ways: by checking word definitions, by analysis of their usage in texts, or by using formal mathematical models. Formally measuring word similarities enables systems to understand and process human languages more effectively. Dissimilar words have lower word similarity scores. Listing 4-8 is a Python demo of this important NLP data preprocessing step.

Listing 4-8. Word Similarity Measurements Using NLTK

```
# Install nltk if you have not done it already.
!pip install nltk # The code output not included here.
# Load the reqi
import nltk
from nltk.corpus import wordnet

# Download that WordNet.
nltk.download('wordnet')

def word_similarity(word1, word2):
    # Get synsets for both words
    synsets1 = wordnet.synsets(word1)
    synsets2 = wordnet.synsets(word2)

    if not synsets1 or not synsets2:
        return "One or both words are not found in WordNet."

    # Calculate similarity with all pairs of synsets.
    similarities = []
    for syn1 in synsets1:
        for syn2 in synsets2:
            similarity = syn1.wup_similarity(syn2)
```

```python
    if similarity is not None:
        similarities.append(similarity)

# Return the highest similarity score.
return max(similarities, default="No similarity score
available")
# Example 1.
sample_word1 = "dog"
sample_word2 = "puppy"

similarity_score = word_similarity(sample_word1, sample_word2)
print(f"Similarity score between '{ sample_word1 }' and '{
sample_word2}': {similarity_score}")
OUTPUT 1:
Similarity score between 'dog' and 'puppy': 0.896551724137931
# Example 2.
sample_word1 = "Canon"
sample_word2 = "House"

similarity_score = word_similarity(sample_word1, sample_word2)
print(f"Similarity score between '{ sample_word1 }' and '{
sample_word2 }': {similarity_score}")
OUTPUT 2:
Similarity score between 'Canon' and 'House':
0.42857142857142855
```

4.2.9 Word Sense Disambiguation Using NLTK

You have seen that many words have multiple meanings depending on the context in which it is being used. Word sense disambiguation (WSD) is a technique to find out which meaning of a word is being used in a sentence. WSD figures out which meaning is the best fit depending on the context.

Let's figure it out with a simple example word: "bat." It can refer to a flying mammal or a piece of sports kit used in baseball or cricket. WSD determines whether "bat" means the animal or the equipment depending on the context of the sentence. You might be wondering how a machine (read) computer does all of it. The machine's NLP software simply analyzes the surrounding words to clarify the intended meaning of a specific word under consideration (see Figure 4-4).

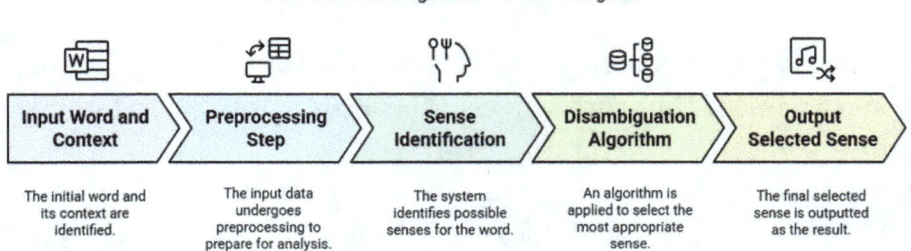

Figure 4-4. *WSD using NLTK*

NLTK again leverages its lexical resource WordNet for WSD. WordNet is equipped with definitions and relationships between words. NLTK uses algorithms that automatically choose the correct meaning of a word depending on its context in a sentence. Listing 4-9 is the code demo for this important NLP data preprocessing operation.

Listing 4-9. WSD Using NLTK

```
# Install nltk if you have not done it already.
!pip install nltk # The code output not included here.
import nltk
from nltk.corpus import wordnet
from nltk.wsd import lesk
```

```
# Ensure required data is downloaded
nltk.download('wordnet')
nltk.download('omw-1.4')

def disambiguate_word(sentence, word):
    # Apply the Lesk algorithm to disambiguate the word
    sense = lesk(nltk.word_tokenize(sentence), word)
    return sense.definition() if sense else "No sense found."
# Example usage 1
sentence = "He uses a bank to deposit his lard earned money."
word_for_WSD1 = "bank"

sense_definition1 = disambiguate_word(sentence, word_for_WSD1)
print(f"Sense of '{word}': {sense_definition1}")
Sense of 'bank': a container (usually with a slot in the top) for keeping money at home
# Example usage 2
sentence = "The US baseball players usually show great skills with bat."
word_for_WSD2 = "bat"

sense_definition2 = disambiguate_word(sentence, word_for_WSD2)
print(f"Sense of '{word}': {sense_definition2}")
OUTPUT:
Sense of 'bank': strike with, or as if with a baseball bat
```

4.2.10 Keyword Extraction with NLTK

In simple words, keyword extraction is getting the most important words or phrases from a given text corpus. Keywords capture the crux of the content. These keywords can be used for other NLP tasks like text summarization, search engine optimization, and retrieval of important information. Let's take an example from our own tech world. In a technical document about machine learning or deep learning models, keywords might include "neural networks," "deep learning," "machine learning," "model accuracy," and "model accuracy." These are keywords that describe the core concepts and methods described in the tech document.

Keyword extraction is done by analyzing the frequency and relevance of words in a given text document. Standard techniques like term frequency (TF) and term frequency–inverse document frequency (TF–IDF) are often used for this purpose. These techniques highlight the significant words for the text under consideration. As such, simple tokenization of the text and computing word frequencies can identify the most important words. Combining it with advanced techniques like TF–IDF makes more sophisticated keyword extraction possible. This chapter only uses these techniques (as used in NLTK), but we formally discuss them in later chapters. Listing 4-10 is the code demo of this important NLP data preprocessing task.

Listing 4-10. Construct a Keyword Extractor Using NLTK

```
# Install nltk if you have not done it already.
!pip install nltk # The code output not included here.
# Import necessary libraries/ resources.
import nltk
from nltk.tokenize import word_tokenize
from nltk.probability import FreqDist
from nltk.corpus import stopwords

# Download required data.
nltk.download('punkt')
nltk.download('stopwords')

def keywords_extractor(text, num_keywords=3):
    # Tokenize the input text.
    tokens = word_tokenize(text.lower())  # lowercase for consistency.
```

```python
# Eliminate stopwords and non-alphanumeric tokens.
stop_words = set(stopwords.words('english'))
filtered_tokens = [word for word in tokens if word.
isalnum() and word not in stop_words]

# Calculate the input word frequencies.
freq_dist = FreqDist(filtered_tokens)

# Extract the most frequent keywords.
keywords = [word for word, _ in freq_dist.most_common(num_
keywords)]
    return keywords
# Example usage
sample_text = "The bank is on the city river bank. State Bank
is the best bank in the city."

keywords = keywords_extractor(sample_text)
print(f"Extracted keywords: {keywords}")
OUTPUT:
Extracted keywords: ['bank', 'city', 'river']
```

This section offered an inclusive overview of some essential NLP data preprocessing tasks using NLTK. We explored common techniques such as tokenization, stop word removal, stemming, lemmatization, and sentence segmentation to prepare text for analysis. Adding to these basic NLP techniques, we also covered WFD, synonym and antonym detection, word similarity calculation, and WSD. Finally, we looked into keyword extraction techniques to get key terms out of any input text document. All the techniques discussed in this section are vital for transforming raw text information into actionable insights and improving the effectiveness of NLP applications. The next section discusses some important NLP techniques beyond the basic preprocessing tasks.

4.3 NLP Tasks Beyond Preprocessing

This section delves into some new NLP techniques. We start with part-of-speech/POS tagging, a process used to label every word in a given input sentence considering its grammatical role, which can be a noun, verb, or adjective. Next, let's explore named entity recognition, which identifies proper names (in a given sentence), like the name of a person, organizations, and locations, including cities and countries. We then move to a relatively known term, *text classification*, which categorizes text into predefined labels, such as categorizing emails as junk or legitimate. Following that, we discuss *sentiment analysis*, which determines the emotional tone behind words; examples are classifying tweets about a famous movie into labels: positive or negative.

Finally, we cover a bit of advanced NLP applications like *language translation*, which adapts text from one language to another. The next process, *text summarization*, condenses long text documents into their condensed versions while preserving key information. *Word cloud generation* is a visual illustration of word frequency. It highlights the most important terms in a given text document. All these techniques together allow a deeper and more nuanced understanding of the input text information. For each of these techniques, we give a hands-on, NLTK-based code demonstration, which is the primary focus of this book.

4.3.1 Part-of-Speech Tagging

Part-of-speech, or POS, tagging labels each word in a given input sentence with its grammatical role, such as noun, verb, adjective, or adverb. POS tagging helps in a proper understanding of the structure and meaning of an input text by analyzing the role of each word within the sentence. For example, in the sentence "She reads a book every night," POS tagging assigns "she" as a pronoun, "reads" as a verb, "a" as an article, "book" as a noun, and "every" as an adjective. This helps us (read computers)

CHAPTER 4   NLP DATA PREPROCESSING TASKS WITH NLTK

recognize the role of each word in the sentence. In simple words, by accurately tagging words, POS tagging offers insights into the syntax and grammar of a given input text document. For example, POS tagging helps in disambiguating words having diverse meanings based on their role in a sentence; POS tagging also lays the foundation for more advanced tasks like parsing, sentiment analysis, and named entity recognition (see Figure 4-5). Consider the following two sample sentences.

- Sample 1: "The light in the room was bright."
- Sample 2: "She carried a light bag."

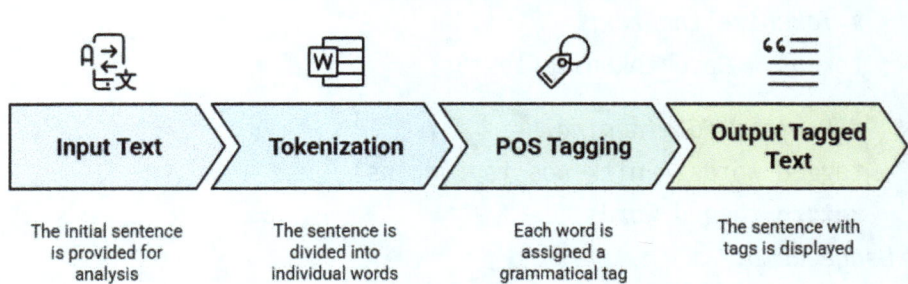

Figure 4-5. POS tagging using NLTK

In sample 1, "light" serves as a noun (meaning something that illuminates), while in sample 2, "light" acts as an adjective (meaning something that is not heavy). This difference is made clear to machines (running NLP software) through POS tagging, which helps them to realize the word's role in different contexts. POS tagging is the groundwork for more advanced NLP tasks that include parsing, sentiment analysis, and NER. Listing 4-11 is a demo that uses NLTK to accomplish POS tagging.

Listing 4-11. POS Tagging Using NLTK

```
# Install nltk if you have not done it already.
!pip install nltk # The code output not included here.
# Import libraries.
import nltk
from nltk.tokenize import word_tokenize

# Download required data.
nltk.download('punkt')
nltk.download('averaged_perceptron_tagger')

def pos_tagging(text):
    # Tokenize the text.
    tokens = word_tokenize(text)

    # Perform POS tagging.
    tagged_words = nltk.pos_tag(tokens)
    return tagged_words
# Usage demo.
sentence = "The school girl quickly ran across the park to catch the ice cream truck."
tagged_sentence = pos_tagging(sentence)
print(f"POS tagging result: {tagged_sentence}")
OUTPUT:
POS tagging result: [('The', 'DT'), ('school', 'NN'), ('girl', 'NN'), ('quickly', 'RB'), ('ran', 'VBD'), ('across', 'IN'), ('the', 'DT'), ('park', 'NN'), ('to', 'TO'), ('catch', 'VB'), ('the', 'DT'), ('ice', 'NN'), ('cream', 'NN'), ('truck', 'NN'), ('.', '.')]
```

In Listing 4-11 output, notice that several POS tags (coded as 'DT', 'NN', etc.) are used. For your convenience, Table 4-1 lists the meaning of each tag used in the output.

CHAPTER 4　NLP DATA PREPROCESSING TASKS WITH NLTK

Table 4-1. *POS Tags Used in the Output of Code Demo*

POS Tags	Meaning
DT:	Determiner - Introduces a noun.
NN:	Noun, Singular or Mass - Represents a person, place, thing, or idea.
RB:	Adverb - Modifies a verb, adjective, or adverb.
VBD:	Verb, Past Tense - Indicates an action completed in the past.
IN:	Preposition or Subordinating Conjunction - Shows relationships between words or introduces a clause.
TO:	To - Used before a verb to form the infinitive or indicate direction/purpose.
VB:	Verb, Base Form - The base form of a verb.
.:	Punctuation - Marks the end of a sentence.

4.3.2 Named Entity Recognition with NLTK

Named entity recognition (NER) is a NLP technique that involves finding and categorizing explicit entities or objects within a given input text document (see Figure 4-6). These named entities can be an individual person, establishments, time or date expressions, locations, quantity expressions, money, percentages, and alike. NER benefits in extracting meaningful information from unstructured text. For example, consider the following sentence.

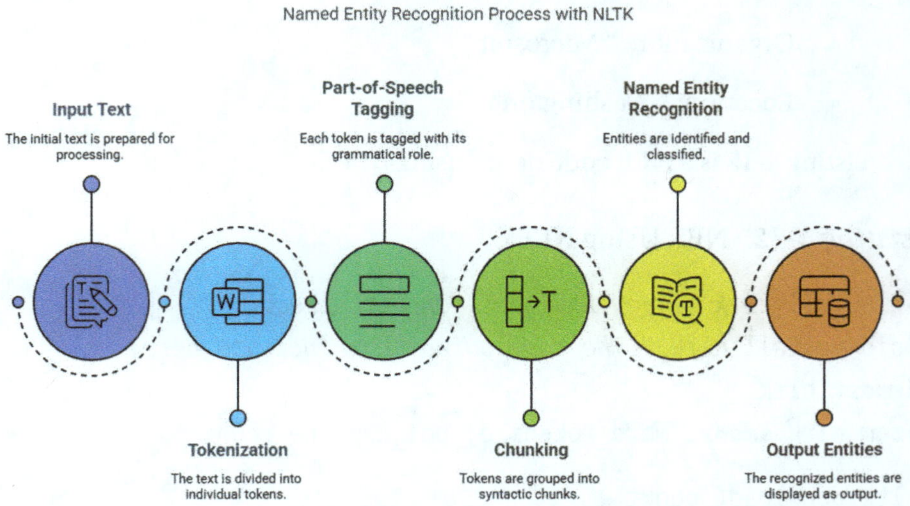

Figure 4-6. *Named entity recognition with NLTK*

CHAPTER 4   NLP DATA PREPROCESSING TASKS WITH NLTK

For example, let's consider the sentence: "Dr. Jonson attended the University of London and now resides in Liverpool." Here, the named entities on a high level are as follows.

- Person: "Dr. Johnson"
- Organization: "University of London"
- Location: "Liverpool"

One direct usage of NER is it can be upfront used by different NLP applications such as chatbots. Suppose any user asks the chatbot, "Which university did Dr. Jonson attend?" It has a ready answer by virtue of extracted entities. This way, NER makes it easier to analyze and categorize the content.

Let's take one more example as follows.

Consider the sentence: "Bill Gates founded Microsoft and resides in Washington."

In this sentence, the NER can tell us the high-level entities as follows.

- Person: " Bill Gates "
- Organization: " Microsoft "
- Location: " Washington"

Listing 4-12 is a NER code demo using NLTK.

Listing 4-12. NER Using NLTK

```
# Install nltk if you have not done it already.
!pip install nltk # The code output not included here.
import nltk
from nltk import word_tokenize, pos_tag, ne_chunk

nltk.download('punkt')
nltk.download('maxent_ne_chunker')
nltk.download('words')
```

```python
def named_entity_extractor(text):
    # Tokenize.
    tokens = word_tokenize(text)

    # Perform POS tagging first.
    pos_tags = pos_tag(tokens)

    # Apply NER.
    named_entities = ne_chunk(pos_tags)
    return named_entities
# Sample usage 1.
sample_sentence1 = "Bill Gates founded Microsoft and resides in Washington."
ner_result = named_entity_extractor(sample_sentence1)
print(f"NER result: {ner_result}")
```
OUTPUT1:
NER result: (S
  (PERSON Bill/NNP)
  (PERSON Gates/NNP)
  founded/VBD
  (PERSON Microsoft/NNP)
  and/CC
  resides/NNS
  in/IN
  (GPE Washington/NNP)
  ./.)
```python
# Sample usage 2.
sample_sentence2 = "Dr. Jonson attended University of London and now resides in Liverpool."
ner_result = named_entity_extractor(sample_sentence2)
print(f"NER result: {ner_result}")
```

CHAPTER 4   NLP DATA PREPROCESSING TASKS WITH NLTK

```
OUTPUT2:
NER result: (S
  Dr./NNP
  (PERSON Jonson/NNP)
  attended/VBD
  (ORGANIZATION University/NNP)
  of/IN
  (GPE London/NNP)
  and/CC
  now/RB
  resides/VBZ
  in/IN
  (GPE Liverpool/NNP)
  ./.)
```

You can decode NER codes, including NNP, VBD, NNP, and VBZ, on the Penn Treebank Project webpage at www.ling.upenn.edu/courses/Fall_2003/ling001/penn_treebank_pos.html.

4.4 Some Useful Functionalities Outside NLTK

This section explores a few prevailing NLP text processing functionalities that go outside the capabilities currently available in NLTK. These important functionalities include word cloud generation, language translation, and text summarization.

Word Cloud Generation is a NLP tool that highlights the words with the most frequencies from the input text. Just a glance at word cloud can reveal the key themes discussed in the input text document. Language translation (or machine translation) allows the automatic conversion of input text from one language to another to make content accessible to a global audience. Text Summarization condenses large volumes of text into concise summaries that preserve the main ideas. These functionalities further enhance the utility of NLP applications to the international audience. Next, we begin by explaining word cloud generation with a code demo.

4.4.1 Word Cloud Generation

Word cloud generation (WSG) is an influential visual tool that is used in text analytics. It represents the most frequent words available within an input text document. Words appearing more frequently are displayed in larger, bolder fonts. This type of visual representation of words helps to quickly categorize the central themes and key terms. Word clouds are often used by the NLP community for exploratory data analysis (EDA). Word clouds are predominantly used in summarizing large text documents. However, word clouds do not reveal the context in which words appear. Despite this major limitation, word clouds are popular for initial text exploration. Listing 4-13 contains a sample code of WSG.

Listing 4-13. Word Cloud Generation (currently not available in NLTK)

```
# Install wordcloud if you do not have it already.
!pip install wordcloud
# Import required libraries.
import nltk
from wordcloud import WordCloud
import matplotlib.pyplot as plt
from nltk.corpus import stopwords
from nltk.tokenize import word_tokenize

# Download data.
```

```python
nltk.download('punkt')
nltk.download('stopwords')

# Write necessary functions.
def text_preprocessor(text):
    """
    Tokenize and remove stopwords.

    Args:
    text (str): The imput text.

    Returns:
    str: The preprocessed text.
    """
    tokens = word_tokenize(text.lower())
    filtered_tokens = [word for word in tokens if word.isalnum() and word not in stopwords.words('english')]
    return ' '.join(filtered_tokens)

def word_cloud_generator(text):
    """
    Generate and display word cloud.

    Args:
    text (str): The preprocessed text.
    """
    wordcloud = WordCloud(width=800, height=400, background_color='white').generate(text)

    plt.figure(figsize=(8, 5))
    plt.imshow(wordcloud, interpolation='bilinear')
    plt.axis('off')
```

```python
    plt.show()
def main():
    # Sample text document.
    sample_text = """
    Word Cloud Generation (WSG) is an influential visual tool
    that is used in text analytics.
    It represents the most frequent words available within an
    input text document.
    Words appearing more frequently are displayed in larger,
    bolder fonts.
    This type of visual representation of words helps to
    quickly categorize the central themes and key terms.
    Word clouds are often used by the NLP community for
    exploratory data analysis EDA).
    Word clouds are predominantly used in summarizing large
    texts documents.
    However, word clouds do not reveal the context in which
    words appear.
    Despite this major limitation, word clouds are popularly
    for initial text exploration (EDA).
    """

    # Preprocess the sample text.
    pre_processed_text = text_preprocessor(sample_text)

    # Generate the word cloud.
    word_cloud_generator(pre_processed_text)

if __name__ == "__main__":
    main()
```

CHAPTER 4   NLP DATA PREPROCESSING TASKS WITH NLTK

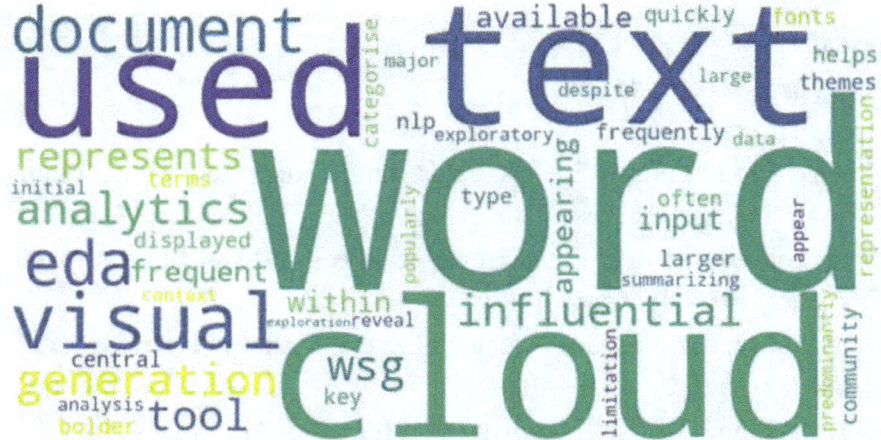

Figure 4-7. *The Word Cloud generated by the code of Listing 4-13*

4.4.2 Language Translation

Language translation (or machine translation) is useful in applications like online translation services, multilingual customer support, and global content accessibility. Further, machine translation supports international business operations, permits cross-cultural exchanges, and gives broad access to digital content. In machine translation, accurate interpretation of the original text is critical to replicate the nuances of the original text in the target language.

Listing 4-14 uses the translator library for translating the input text because NLTK does not have this functionality.

Listing 4-14. Creating a Machine Translator with Translator Library

```
# Install translate if you have not done it already.
!pip install translate
from translate import Translator

def text_translator(text, from_lang='en', to_lang='es'):
    # Initialize the Translator.
```

142

```
    translator = Translator(from_lang=from_lang,
    to_lang=to_lang)

    # Translate the text from English to Spanish.
    translated_text = translator.translate(text)
    return translated_text
# Example usage 1.
Original_input_text = "Hello, how are you today? I hope you're doing well."
translated_text = text_translator(Original_input_text)
print(f"Original input text: {Original_input_text}")
print(f"Translated text: {translated_text}")
OUTPUT1:
Original input text: Hello, how are you today? I hope you're doing well.
Translated text: Hola, ¿cómo estás hoy? Espero que lo estés haciendo bien.
# Example usage 2.
original_text = "Hello, we won the baseball game today and received the prize money."
translated_text = text_translator(original_text)
print(f"Original text: {original_text}")
print(f"Translated text: {translated_text}")
OUTPUT2:
Original text: Hello, we won the baseball game today and received the prize money.
Translated text: Hola, hoy ganamos el juego de béisbol y recibimos el dinero del premio.
```

By the way, if you happen to know Spanish, you can judge the quality of this translation. Doing this, you may discover that probably the Spanish text in both the cases in the demo is understandable, but improvements are still possible for better clarity and correctness.

4.4.3 Text Summarization

Text summarization is predominantly useful in quickly understanding the main points of long articles, reports, or documents without reading them completely. It's a key skill in the fields like journalism, legal work, and academic research. Summarization has increasingly become central in automated systems like chatbots, recommendation engines, and news aggregators. In all these applications, delivering concise information is critical. Summarization is challenged to accurately identify the most relevant content to keep the text summary contextually appropriate. Listing 4-15 is a code demonstration of automatic text summarization.

Listing 4-15. Text Summarization (currently not available in NLTK)

```
# Install sumy if you have not done already.
!pip install sumy
# Import the required libraries.
from sumy.parsers.plaintext import PlaintextParser
from sumy.nlp.tokenizers import Tokenizer
from sumy.summarizers.lsa import LsaSummarizer

def text_summarizer(text, sentence_count=2):
    """
    Summarize the input text using the LSA algorithm.

    Args:
    text (str): The input text.
    sentence_count (int): The count of sentences to include in
    the final summary.

    Returns:
    str: The summarized text.
    """
```

```python
    parser = PlaintextParser.from_string(text, Tokenizer
    ("english"))
    summarizer = LsaSummarizer()
    summary = summarizer(parser.document, sentence_count)

    return ' '.join(str(sentence) for sentence in summary)
def main():
    # Sample text
    sample_text = """ Text summarization is predominantly useful in quickly understanding the main points of long articles, reports, or documents without reading them in completely. Text summarization is a key skill in the fields like journalism, legal work, and academic research. Summarization has increasingly become central in automated systems like chatbots, recommendation engines, and news aggregators. In all these applications delivering concise information is critical. Summarization challenged to accurately identifying the most relevant content to keep the text summary contextually appropriate. Below is code demonstration of automatic text summarization.

    """

    # Generate the summary
    summary = text_summarizer(sample_text)

    print("Original Text:")
    print(sample_text)
    print("\nSummary:")
    print(summary)

if __name__ == "__main__":
    main()
```

CHAPTER 4   NLP DATA PREPROCESSING TASKS WITH NLTK

OUTPUT:

Original Text:
 Text summarization is predominantly useful in quickly understanding the main points of long articles, reports, or documents without reading them in completely. Text summarization is a key skill in the fields like journalism, legal work, and academic research. Summarization has increasingly become central in automated systems like chatbots, recommendation engines, and news aggregators. In all these applications delivering concise information is critical. Summarization challenged to accurately identifying the most relevant content to keep the text summary contextually appropriate. Below is code demonstration of automatic text summarization.

Summary:
Text summarization is a key skill in the fields like journalism, legal work, and academic research. Summarization has increasingly become central in automated systems like chatbots, recommendation engines, and news aggregators.

Note   There are many other important Python libraries, including spaCy, Gensim, and several others. Due to the space limits of individual chapters, we discuss them with the relevant NLP tasks that can be best solved by these libraries. Stay tuned!

4.5 Chapter Recap

This chapter provided an in-depth exploration of various NLP techniques. Preprocessing of the raw text data is at the core of NLP to make the raw data usable for further analysis. We began the chapter by text tokenization, which involves breaking down text into manageable pieces. Following this, we tackled the removal of irrelevant words, called stop words, to keep the focus on more meaningful content. We also explored stemming and lemmatization. These are the techniques that simplify words to their root forms so that similar words can be grouped together to improve the analysis process.

Sentence segmentation (along with finding WFD) permitted us to better understand the input text structures. Finding word frequencies also helped us to categorize key terms within a document. Further, we explored how to detect synonyms and antonyms that are crucial for NLP tasks like paraphrasing. We then progressed to more complex tasks like calculating word similarity. Finding similarity allows machines to compare and contrast different pieces of text. Exploring WSD helped us learn how to resolve word ambiguities. WSD finds the correct meaning of a word based on its context in the input text. Keyword extraction was another key focus. It enabled us to focus on the most relevant terms within a large text corpus.

Moving beyond text preprocessing, we explored tasks such as POS tagging and NER. POS tagging assigns the proper grammatical role to every word, while NER categorizes entities like names, organizations, and locations within a given text corpus. These tools make possible deeper text analysis and a better understanding of text.

In the last leg of this chapter, we explored some important functionalities outside of NLTK. It included word cloud generation for exploratory analysis of text data. We also studied machine translation for content globalization and text summarization for condensing lengthy texts into concise summaries. The aim of this chapter was to equip you with NLP techniques and provide a comprehensive toolkit for handling and analyzing text data.

CHAPTER 4   NLP DATA PREPROCESSING TASKS WITH NLTK

Code demos are given with every technique discussed in this chapter so that you can directly use the given code for a variety of NLP applications, ranging from academic research to hardcore business applications.

4.6 Exercises

1. In any text analytics task, such as text classification or Twitter sentiment analysis (about a product or a movie), compare the results with and without the stop words. If you are new to text classification or Twitter sentiment analysis, you can get the tutorials on Kaggle.com or with many other articles. These text analytics projects are covered in the upcoming chapters.

2. Generate the text summary and word cloud for a relatively large text document (minimum two pages) and observe if all the prominent words highlighted in the word cloud appear in your summary.

CHAPTER 5

Lexical Analysis

5.1 Why You Should Read This Chapter

For machines to understand and structure the input text, breaking down text into smaller units and categorizing them is an essential process. Lexical analysis exactly does this; it involves breaking down the input text into its utmost basic units like words, morphemes, and tokens. Morphemes are the smallest units of meaning in a language. They represent distinct words, like "dog" or "read." Morphemes also includes parts of words, such as prefixes ("un-"), suffixes ("-ing"), or roots ("tele-" as in "telecommunication"). Morphemes are fundamental blocks in the structure of any language.

Lexical analysis transforms unstructured text into structured data, which is essential for advanced NLP tasks like parsing, semantic analysis, and information retrieval. Accurate identification and categorization of language components using lexical analysis helps in improving the efficiency and effectiveness of several real-life NLP applications. Lexical analysis ultimately enables computers to better understand what humans speak and write.

5.2 Morphological Analysis Revisited

We discussed morphological analysis at the start of this chapter. By learning how morphemes combine to form words, we can understand language patterns and enable several language processing tasks like stemming and lemmatization. Let's try to understand morphemes with the help of the following three familiar words.

- The word "unhappiness" has three morphemes. The prefix "un-" (meaning "not"), the root word "happy" (meaning "joyful"), and finally the suffix "-ness" (meaning "state of being"). Combined, they form the word "unhappiness."

- "Replayed" has two morphemes. The prefix "re-" (meaning "again") and the root word "played" (meaning "engaged in a game or activity").

- Similarly, the word "cats" contains two morphemes: the base noun "cat" and the suffix "-s" (representing the plural form).

By analyzing the structure of words, we can efficiently extract meaningful features and develop more effective language models.

There are two major categories of morphemes.

- **Free morphemes** can exist and stand alone as independent words with meaning. Examples are "book," "run," and "happy." These morphemes convey meaning even if they are not attached to any other word.

- **Bound morphemes** must be attached to a free morpheme to have meaning. Consider our familiar example, "unhappy." Here, prefixes like "un-" in the base word "happy" modify the meaning of the base word. This way, bound morphemes further add nuances or grammatical context to words.

Free morphemes represent the basic vocabulary units of a language, while bound morphemes, such as prefixes, suffixes, and infixes, help us to understand how words are modified and related to each other. Understanding bound morphemes is helpful in stemming and lemmatization because both processes aim to reduce words to their base or root forms. This knowledge can be useful in search engines, which need different forms of a word (e.g., "run," "running," "runner") to be recognized as related. Additionally, knowledge morphemes allow more accurate matching of words and phrases, even if they appear in different forms. For example, knowing that the following three words, "running," "runner," and "runs," come from the same morpheme (base), "run" can improve the grouping of connected texts.

Reading this, you can imagine that morphological analysis can find its uses in almost all the NLP techniques that we studied in Chapter 4. For example, morphological analysis is useful in stemming and lemmatization, part-of-speech (POS) tagging, named entity recognition (NER), machine translation, text normalization, sentiment analysis, and several other NLP tasks.

The code demos for the morphological analysis are presented in the following sections, which contain the related techniques.

5.3 Tokenization Revisited

We studied basic tokenization in the previous chapter. We also studied how the word_tokenize() and sent_tokenize() methods from Python's NLTK library can be utilized to break a larger text corpus (or text document) at sentence level or word level, respectively. Gensim and Keras are a two other Python open source libraries that are available for tokenization.

5.3.1 Code Demos

Let's discuss some other tokenizers: whitespace, dictionary-based, rule-based, punctuation-based, tweets, and multi-word expressions. Each tokenizer serves a different purpose. We provide a bit of theory about each tokenizer in the code itself.

5.3.1.1 Whitespace Tokenization

Whitespace tokenization breaks a text document based on spaces and line breaks. It considers every section of input text divided by whitespace as a distinct token. Whitespace tokenization serves as a quick way to break down text into words or phrases. This method is especially used for processing structured data, including log files or code, in which whitespaces have specific meanings.

Whitespace tokenization is less worried about the grammatical structure of input texts. Its main focus is on the visual separation of text. Whitespace tokenization is used in the NLP tasks where sentence boundaries are not as important. Examples include keyword extraction or simple text preprocessing, where the focus is on quickly splitting the input text into smaller and more manageable tasks. The following is a code demo.

```python
from nltk.tokenize import WhitespaceTokenizer
from typing import List

def whitespace_tokenizer(text: str) -> List[str]:
    """
    Tokenizes the input text based on whitespace characters
    (spaces, tabs, newlines).

    Parameters:
    -----------
    text : str
        The input text to be tokenized.
```

```
    Returns:
    --------
    List[str]
        A list of tokens.
    """
    tokenizer = WhitespaceTokenizer()
    tokens = tokenizer.tokenize(text)
    return tokens
# Example usage
if __name__ == "__main__":
    sample_text = """Whitespace tokenization breaks a text
    document based on spaces and line breaks. It considers every
    section of input text divided by whitespace as a distinct
    token. Whitespace tokenization serves as a quick way to break
    down text into words or phrases. This method is especially
    used for processing structured data including logs files or
    code, in which whitespaces have specific meaning.
    """
    tokens = whitespace_tokenize(sample_text)
    print("Tokens:", tokens)
OUTPUT:
Tokens: ['Whitespace', 'tokenization', 'breaks', 'a', 'text',
'document', 'based', 'on', 'spaces', 'and', 'line', 'breaks.',
'It', 'considers', 'every', 'section', 'of', 'input', 'text',
'divided', 'by', 'whitespace', 'as', 'a', 'distinct', 'token.',
'Whitespace', 'tokenization', 'serves', 'as', 'a', 'quick',
'way', 'to', 'break', 'down', 'text', 'into', 'words', 'or',
'phrases.', 'This', 'method', 'is', 'especially', 'used',
'for', 'processing', 'structured', 'data', 'including', 'logs',
'files', 'or', 'code,', 'in', 'which', 'whitespaces', 'have',
'specific', 'meaning.']
```

5.3.1.2 Dictionary-Based Tokenization

In dictionary-based tokenization, the tokens are based on the words that are already available in the dictionary. In case the token is not found in the dictionary, special rules are applied to tokenize it.

Dictionary-based tokenization is used where identifying and extracting predefined phrases or terms is more important. It ensures use in domain-specific texts, where consistency in recognizing specific tokens is more important than simple tokens. This method is particularly useful in NLP tasks like medical text processing and parsing of legal documents, where precise recognition of terms is critical.

```
import nltk
from nltk.tokenize import MWETokenizer
from typing import List

# NLTK data
nltk.download('punkt')

def dictionary_based_tokenizer(text: str, dictionary: List[List[str]]) -> List[str]:
    """
    Tokenizes the input text based on a predefined dictionary.

    Parameters:
    -----------
    text : str
        The input text string.
    dictionary : List[List[str]]
        A list of expressions, recognized as single tokens.

    Returns:
    --------
    List[str]
        A list of tokens.
```

```python
    """
    # The '_' argument joinS the words with an underscore (_).
    tokenizer = MWETokenizer(dictionary, separator='_')
    tokens = tokenizer.tokenize(nltk.word_tokenize(text))
    return tokens
# Example usage
if __name__ == "__main__":
    # Define a dictionary of multi-word expressions
    dictionary = [
        ['heart', 'attack'],
        ['machine', 'learning'],
        ['New', 'York'],
    ]

    sample_text = "John suffered a messive heart attack. He was enrolled in machine learning in Washington DC."
    tokens = dictionary_based_tokenizer(sample_text, dictionary)
    print("Tokens:", tokens)
```
OUTPUT:
Tokens: ['John', 'suffered', 'a', 'messive', 'heart_attack', '.', 'He', 'was', 'enrolled', 'in', 'machine_learning', 'in', 'Washington', 'DC', '.']

Note You may notice in the output that 'heart attack' appears as 'heart_attack' and 'machine learning' as 'machine_learning'.

5.3.1.3 Rule-based Tokenization

Rule-based tokenization is particularly useful in text preprocessing tasks where precise structures or formats must be preserved. Regular expressions can be used to fine-tune to handle various linguistic nuances or specific text formats. The following uses RegexpTokenizer from the NLTK along with required regular expression patterns.

```
from nltk.tokenize import RegexpTokenizer
from typing import List

def rule_based_tokenizer(text: str) -> List[str]:
    """
    Tokenizes the input text based on regular expressions.

    Parameters:
    -----------
    text : str
        The input string.

    Returns:
    --------
    List[str]
        A list of tokens.
    """
    # Define a regular expressions for specific tasks.
    # Rule 1: Match abbreviations like "U.K.", "Dr."
    # Rule 2: Match words including those with apostrophes like "she's", "haven't")
    # Rule 3: Separately match punctuations.
    pattern = r'''(?x)              # Set flag to allow
                                    verbose regex
            (?:[A-Za-z]\.)+         # Abbreviations like U.K.
          | \w+(?:'\w+)?            # Words with apostrophes
```

CHAPTER 5　LEXICAL ANALYSIS

```
        | [^\w\s]              # Separate punctuations
    '''

    # Write a RegexpTokenizer.
    tokenizer = RegexpTokenizer(pattern)

    # Tokenize.
    tokens = tokenizer.tokenize(text)

    return tokens
# Example usage
if __name__ == "__main__":
    sample_text = """Dr. John's a rocket engineer in the U.K.
                    U.A.E. govt. has awarded him many awards."""
    tokens = rule_based_tokenizer(sample_text)
    print("Tokens:", tokens)
OUTPUT:
Tokens: ['Dr', '.', "John's", 'a', 'rocket', 'engineer', 'in',
'the', 'U.K.', 'U.A.E.', 'govt', '.', 'has', 'awarded', 'him',
'many', 'awards', '.']
```

- Please note, in 'Dr', 'U.K.', and 'U.A.E.', the tokenizer has preserved common abbreviations and acronyms and treated them as single tokens.

- 'govt': This is treated as a single token, preserving abbreviations that are commonly used in written English.

- For "John's", the tokenizer preserves the apostrophe and the following 's' as part of the token.

157

5.3.1.4 Punctuation-Based Tokenizer

This tokenizer splits the input text on whitespace and punctuations while retaining the punctuations. Punctuation-based tokenization overcomes issues and provides a meaningful token. Punctuation-based tokenization solves the problem of simply splitting the input text on whitespace. The following is a simple code demo.

```
#import wordpunct_tokenize from nltk.
from nltk.tokenize import wordpunct_tokenize

text = "Mr.Johnson owns three houses in New York on: 22nd street, downton,and 42nd street."

tokens = wordpunct_tokenize(text)
tokens
OUTPUT:
['Mr',
 '.',
 'Johnson',
 'owns',
 'three',
 'houses',
 'in',
 'New',
 'York',
 'on',
 ':',
 '22nd',
 'street',
 ',',
 'downton',
 ',',
```

```
'and',
'42nd',
'street',
'.']
```

> **Note** The output retains punctuation marks as separate tokens. It helps in understanding the sentence structure. The presence of punctuation tokens helps preserve the meaning and context of the original sentence.

5.3.1.5 Tweet Tokenizer

Tweets can be treated as special texts like several others because they have a typical structure. The generic tokenizers may fail to produce feasible tokens when applied to the datasets containing tweets. NLTK has a special tokenizer for such cases. The Tweet tokenizer is a rule-based tokenizer that can be used to remove problematic characters, HTML code, and Twitter handles. This tokenizer can normalize the length of input text by reducing the occurrence of repeated letters. Additionally, the tweet tokenizer is equipped to effectively deal with hashtags and mentions while preserving their significance. The tokenizer can also effectively manage emoticons and special symbols. Note that hashtags and emoticons are common in tweets. The following is a short but reusable code demo using this tokenizer.

```
from nltk.tokenize import TweetTokenizer

def punctuation_based_tokenizer(text):
    """
    Tokenizes the input text using NLTK's TweetTokenizer, which
    handles punctuation and special characters
```

commonly found in tweets.

Parameters:
text (str): The input text to be tokenized.

Returns:
list: A list of tokens, including words and punctuation marks.
```
"""
# Initialize the tokenizer.
tokenizer = TweetTokenizer()

# Tokenize.
tokens = tokenizer.tokenize(text)

return tokens
# Example usage
if __name__ == "__main__":
    Sample_text = """"Mr.Johnson owns three houses in New York! 22nd street, downton,and 42nd street." #marvelous @Rita"
    """
    tokens = punctuation_based_tokenizer(Sample_text)
    print(tokens)
OUTPUT:
['Mr.Johnson', 'owns', 'three', 'houses', 'in', 'New', 'York', '!', '22nd', 'street', ',', 'downton', ',', 'and', '42nd', 'street', '.', '"', '#marvelous', '@Rita', '"']
```

- Please note: Punctuation marks like !, ,, and . are preserved as separate tokens.
- Hashtags (#marvelous) and mentions (@Rita) are preserved as distinct tokens.

- Mr.Johnson is kept intact.
- These are examples to put an emphasis on the fact that TweetTokenizer is specifically tailored for the nuances of social media text.

5.3.1.6 MWE Tokenizer

A multi-word expression (MWE) tokenizer handles sequences of words that act as a single unit in the input text. Examples include idiomatic phrases, compound nouns, and fixed expressions that can be misunderstood if tokenized into separate words. Treating these expressions as single tokens preserves their intended meaning and helps in accurate text analysis. Listing 5-1 is a text demo of this tokenizer with a reusable function.

Listing 5-1. Program to Demonstrate MWE Tokenizer

```
from nltk.tokenize import MWETokenizer

def mwe_tokenizer(text, mwe_list):
    """
    Tokenizes while preserving indicated multi-word
    expressions (MWEs).

    Parameters:
    text (str): The input text.
    mwe_list (list of tuples): The list of MWEs to be treated
    as single tokens.

    Returns:
    list: A list of tokens with MWEs preserved as
    single tokens.
    """
```

CHAPTER 5 LEXICAL ANALYSIS

```python
    # Initialize the MWETokenizer with the provided MWEs
    tokenizer = MWETokenizer(mwe_list)

    # Tokenize the input text
    tokens = tokenizer.tokenize(text.split())

    return tokens
if __name__ == "__main__":
    # Define multi-word expressions.
    mwe_list = [
        ('M', 'W', 'E'),
        ('Multi', 'Word', 'Tokenizer'),
        ('Natural', 'Language', 'Processing'),
        ('pre', 'processing')
    ]
 # Input text
text = "M W E Tokenizer is an advanced tokenizer in Natural
Language Processing pre processing tasks"

# Perform tokenization with MWEs
tokenized = mwe_tokenizer(text, mwe_list)

# Print the results.
print(tokenized)
OUTPUT:
['M_W_E', 'Tokenizer', 'is', 'an', 'advanced', 'tokenizer',
'in', 'Natural_Language_Processing', 'pre_processing', 'tasks']
```

- Note that M_W_E, natural language processing and preprocessing are combined as single units as per the given mwe_list.

- Other task-specific tokenizers include Penn Treebank/ default tokenization, spaCy, Moses, and Subword. They are not covered in this book.

- **Penn Treebank/default tokenization** is used for splitting the input text into tokens depending on linguistic rules. This tokenizer is used for syntactic parsing in the Penn Treebank corpus, which is a popular annotated text dataset that contains syntactically parsed sentences and POS tags. It is primarily used for training and evaluating NLP models.

- The **Spacy tokenizer** is modeled to provide efficient rule-based tokenization that supports intricate linguistic structures. It comes default with the spaCy NLP library.

- The **Moses tokenizer** is used for machine translation tasks. It can handle punctuations and special characters to advance alignment and improve translation quality.

- **Subword tokenization** breaks down words into smaller units or subwords. It's specially made to deal with out-of-vocabulary words.

5.3.1.7 Bag-of-Words Model

The bag-of-words (BoW) model is classically done after tokenization, which is then used to create the BoW representation. The BoW model is one of the primary techniques in NLP. BoW is used to represent text data as a collection of words, regardless of grammar and order of words in the input text. BoW works on the frequency or presence of words in a text. This is the technique it employs to convert text into numerical features that are essential for further analysis. BoW technique treats every word as an individual token and creates a "bag," in which words appearing frequently are considered to be dominating.

BoW and tokenization are related concepts. Tokens are the building blocks for the BoW model. The model then counts the occurrence and frequency of each token across the input text. The BoW model is widely used for many NLP tasks, such as text classification and sentiment analysis. This technique transforms text into a machine-readable numeric format. We demonstrate the BoW model with an example.

Once the tokenization step in text preprocessing is complete, we are left with variable-length sequences of text (each sentence has a different number of words). However, machine learning algorithms need fixed-length vectors of numbers for each sentence in the larger input text. The BoW model overcomes this challenge for us. It counts how many times every word appears in an input text document with multiple sentences (see Figure 5-1). It is like keeping all tokens in a bag where the order of the words is ignored. We were only concerned about whether a word appeared or not.

	the	red	dog	cat	eats	food
1. the red dog	1	1	1	0	0	0
2. cat eats dog	0	0	1	1	1	0
3. dog eats food	0	0	1	0	1	1
4. red cat eats	0	1	0	1	1	0

Figure 5-1. *BoW example*

Figure 5-1 has four inputs sentenced as numbered toward the left. All these four sentences combined only five unique words, as shown in the table heading row. For the first sentence, "the red dog," only the cells corresponding to the words "the," "red," and "dog" have an entry of number '1' each. All the remaining cells have an entry of numeric "0."

We did the same for all the remaining three sentences. Once the BoW process is complete, there are numeric and common length vectors for all the input sentences.

In the first sentence," the red dog," the corresponding numeric vector is [1,1,1,0,0,0], and similarly, for "cat eats dog," there are the corresponding vector [0,0,1,1,1,0]. There are similar vectors for the sentences 3 and 4 as well.

Two things were achieved in this process.

- Text sentences were converted to machine-readable numeric format.

- All vectors were standardized to the exact length as required by machine learning algorithms.

There is linguistic reasoning behind this approach of the BoW model: similar text documents share similar domain-specific vocabularies. For instance, all football-related articles frequently use words like score, pass, and team. Weather reports, on the other hand, use an entirely different set of words, such as rain, sun, and umbrella.

Remember to remove stop words to make it easier to identify similar documents, as stop words are common in almost all documents.

Now, you have come to the end of this section, where we explored several tokenizers and the BoW model to deal with special cases and specific text processing needs. Whitespace tokenization is the most upfront approach, which breaks text based on spaces. It works well for simpler applications. Dictionary-based tokenizers leverage predefined word lists to guarantee accurate token segmentation. Such tokenizers are used to handle known terms and proper nouns. Rule-based tokenizers apply predefined language rules to accommodate various text patterns and exceptions.

Punctuation-based tokenizers list punctuation marks as delimiters. It's useful in cases where text punctuations play a different role. Tweet Tokenizers are designed to handle exclusive aspects of social media tweets that include hashtags, mentions, and emojis. Finally, the MWE tokenizers deal with the challenge of multi-word expressions to ensure that common phrases and idiomatic expressions are treated as single tokens. We closed this section by defining a few more specific tokenizers. Every tokenizer is designed for a specific case. One has to be careful while choosing tokenizers. This choice is dependent on the requirements of the text and the task at hand.

5.4 Stemming Revisited

Tokenization separates text into words, BoW then counts their frequency, and stemming cuts them to root forms—all three steps help to convert raw input text into usable features intended for NLP models. The previous chapter used the Porter stemmer for our code demo. It is one of the earliest stemming algorithms. Porter stemmer has been widely tested and used across various NLP applications for its efficiency and simplicity in breaking words into their root forms. One more reason for this widespread popularity and use is that its stemming rules are well structured to avoid over-stemming and under-stemming. Over-stemming is where different (distinct) words are wrongly reduced to the same stem. Under-stemming is the opposite, where closely related words are not reduced to a desired common stem.

A few other stemmers are available, each designed for a specific purpose or cause. They include the Snowball stemmer and the Lancaster stemmer.

5.4.1 Code Demo for Multiple Stemmers

Let's examine a code demo for each of these stemmers. The code provides a short description and purpose for why it is designed.

5.4.1.1 Snowball or Porter2 Stemmer

The Snowball stemmer is also called the Porter2 Stemmer as it's a modification over the original Porter algorithm. Snowball stemmer has improved rules for greater accuracy and consistency. It can be used across multiple languages, avoiding over- and under-stemming.

The following is a code demo that demonstrates the difference between Porter and Snowball stemmers.

```
# Import necessary libraries
from nltk.tokenize import word_tokenize
from nltk.stem import PorterStemmer, SnowballStemmer
import nltk

# Download usual nltk resources.
nltk.download('punkt')

# Initialize the stemmers.
porter_stemmer = PorterStemmer()
snowball_stemmer = SnowballStemmer("english")

# Create the sample text.
text = "In the summer, the researchers were analyzing various hypotheses about the future development."

# Tokenize the sample text.
tokens = word_tokenize(text)

# Apply stemmers.
porter_tokens = [porter_stemmer.stem(token) for token
```

```
in tokens]
snowball_tokens = [snowball_stemmer.stem(token) for token
in tokens]

# Print results.
print("Original Tokens: ", tokens)
print("Porter Stemmed Tokens: ", porter_tokens)
print("Snowball Stemmed Tokens: ", snowball_tokens)
OUTPUT:
Original Tokens:  ['In', 'the', 'summer', ',', 'the',
'researchers', 'were', 'analyzing', 'various', 'hypotheses',
'about', 'the', 'future', 'development', '.']
Porter Stemmed Tokens:  ['in', 'the', 'summer', ',', 'the',
'research', 'were', 'analyz', 'variou', 'hypothes', 'about',
'the', 'futur', 'develop', '.']
Snowball Stemmed Tokens:  ['in', 'the', 'summer', ',', 'the',
'research', 'were', 'analyz', 'various', 'hypothes', 'about',
'the', 'futur', 'develop', '.']
[nltk_data] Downloading package punkt to C:\Users\Shailendra
[nltk_data]     Kadre\AppData\Roaming\nltk_data...
[nltk_data]   Package punkt is already up-to-date!
```

- Please note: From the original sentence, for the word "various", Snowball stemmer it as "various", which better keeps its meaning. Porter stemmer, on the other hand, reduces the same word as 'variou'.

5.4.1.2 Lancaster Stemmer

The Lancaster stemmer is one of the most aggressive stemming approaches because it works on a set of straightforward, recursive rules to cut the input words to their root forms. It is predominantly useful in NLP applications requiring a high level of generalization. For this purpose, it can sacrifice some precision by over-stemming. Applications like topic modeling focus more on broad themes rather than exact word forms. In such cases, the Lancaster stemmer is applied to generalize terms to their root forms to help identify common topics. Listing 5-2 is its code demo with the Porter stemmer.

Listing 5-2. Program to Demonstrate Lancaster Stemmer

```
import nltk
from nltk.tokenize import word_tokenize
from nltk.stem import PorterStemmer, LancasterStemmer

# Download resources.
nltk.download('punkt')

# Initialize both the stemmers.
porter_stemmer = PorterStemmer()
lancaster_stemmer = LancasterStemmer()

# Sample text.
sample_text = "scientific methodologies used by researchers were aimed at improving accuracy in predictive analytics."

# Tokenize the text
tokens = word_tokenize(sample_text)

# Apply Porter Stemmer.
porter_tokens = [porter_stemmer.stem(token) for token in tokens]

# Apply Lancaster Stemmer.
lancaster_tokens = [lancaster_stemmer.stem(token) for token in tokens]

# Print output.
print("Original Tokens: ", tokens)
```

CHAPTER 5   LEXICAL ANALYSIS

```
print("Porter Stemmed Tokens: ", porter_tokens)
print("Lancaster Stemmed Tokens: ", lancaster_tokens)
OUTPUT:
Original Tokens:  ['The', 'scientific', 'methodologies',
'used', 'by', 'researchers', 'were', 'aimed', 'at',
'improving', 'accuracy', 'in', 'predictive', 'analytics', '.']
Porter Stemmed Tokens:  ['the', 'scientif', 'methodolog',
'use', 'by', 'research', 'were', 'aim', 'at', 'improv',
'accuraci', 'in', 'predict', 'analyt', '.']
Lancaster Stemmed Tokens:  ['the', 'sci', 'methodolog', 'us',
'by', 'research', 'wer', 'aim', 'at', 'improv', 'acc', 'in',
'predict', 'analys', '.']
[nltk_data] Downloading package punkt to C:\Users\Shailendra
[nltk_data]     Kadre\AppData\Roaming\nltk_data...
[nltk_data]   Package punkt is already up-to-date!
```

- 'were' to 'were' only. It preserves more of the original word form.

- At the same time, the Lancaster stemmer applies a more aggressive reduction to other words like "were" to "wer."

- The same difference you can notice for 'accuracy.' Porter stemmer reduces it to 'accuraci' while Lancaster stemmer makes it 'acc.'

So far, we have covered the Porter stemmer, Snowball stemmer, and Lancaster stemmer. Porter stemmer is a foundational tool in NLP, while other more specialized stemmers like the Snowball and Lancaster address specific cases. The Snowball stemmer is an improved version of the Porter stemmer, specially designed for handling a broader range of languages with greater accuracy. In contrast, Lancaster stemmer takes an aggressive stemming approach, which makes it particularly effective in environments requiring a high level of generalization. Each of these stemmers handles a specialized use case. Having said that, let's turn our focus to lemmatization.

5.5 Lemmatization Revisited

As a recap, tokenization breaks down the input text into words (tokens), BoW counts them, and stemming simplifies the tokens to roots. Lemmatization further refines them to accurate base forms for better extraction of NLP features aimed at NLP models. The previous chapter discussed lemmatization and its Python code demo at an introductory level. This section focuses on the major challenges faced by this foundational NLP technique and explores other lemmatizers that may help.

Lemmatization is faced with several challenges. First, it depends heavily on the availability and quality of lexical resources. If the lexical database does not contain specific entries or it has inaccuracies, the lemmatization process may yield incorrect or incomplete results. The second challenge can be polysemy, where a single word has multiple meanings. The lemmatizer must be able to resolve the correct meaning depending on context, which is a challenge without advanced semantic understanding. Additionally, lemmatization is generally a resource-intensive process when compared to stemming. This can be a major challenge when the focus is to improve performance, like in real-time NLP applications that involve large datasets, and processing speed is critical. There are rule-based, dictionary-based, and machine learning–based lemmatizations, each with its own advantages. We do not go into detail here, but you can explore further.

There are many popular lemmatizers available in the NLP domain. These include the WordNet lemmatizer (NLTK), spaCy lemmatizer, TextBlob lemmatizer, Pattern lemmatizer, Stanford NLP lemmatizer, and Gensim lemmatizer. Each stands for a different purpose. The following section delve straights into the code.

5.5.1 Code Demo for Multiple Lemmatizers

Due to space constraints, the code demos are only for spaCy lemmatizer and TextBlob lemmatizer. As usual, we provide a brief explanation and purpose of each of these three lemmatizers in the code demos itself.

5.5.1.1 WordNet Lemmatizer (NLTK)

The spaCy lemmatizer is designed using a combination of rule-based and statistical methods to find the base forms of words. It comes integrated with the spaCy NLP library. This lemmatizer is predominantly effective for high-performance and context-aware lemmatization because of its accuracy in reducing words to their base forms. NER, dependency parsing, advanced text analytics, and large-scale data processing are the NLP applications that especially leverage this lemmatizer. The following is a code demo.

```
# Install spacy if you have not done it already.
!pip install spacy
# Install SpaCy English model.
!python -m spacy download en_core_web_sm
import spacy

# Load the SpaCy English model.
nlp = spacy.load('en_core_web_sm')

# Create Sample text.
text = """The SpaCy lemmatizer is designed using a combination
```

of rule-based and statistical methods
to find the base forms of words. It comes integrated with the
SpaCy NLP library.
"""

# Process the text using SpaCyEnglish model.
doc = nlp(text)

# Get lemmatized forms.
lemmatized_tokens = [(token.text, token.lemma_) **for**
token **in** doc]

# Print the output.
print("\n")
print("OUTPUT")
print("\n")
print("Original Text: ", text)
print("Lemmatized Tokens: ", lemmatized_tokens)

OUTPUT

Original Text: The SpaCy lemmatizer is designed using a combination of rule-based and statistical methods
to find the base forms of words. It comes integrated with the SpaCy NLP library.

Lemmatized Tokens: [('The', 'the'), ('SpaCy', 'SpaCy'), ('lemmatizer', 'lemmatizer'), ('is', 'be'), ('designed', 'design'), ('using', 'use'), ('a', 'a'), ('combination', 'combination'), ('of', 'of'), ('rule', 'rule'), ('-', '-'), ('based', 'base'), ('and', 'and'), ('statistical', 'statistical'), ('methods', 'method'), ('\n', '\n'), ('to', 'to'), ('find', 'find'), ('the', 'the'), ('base', 'base'), ('forms', 'form'), ('of', 'of'), ('words', 'word'), ('.', '.'),

('It', 'it'), ('comes', 'come'), ('integrated', 'integrate'), ('with', 'with'), ('the', 'the'), ('SpaCy', 'SpaCy'), ('NLP', 'NLP'), ('library', 'library'), ('.', '.'), ('\n', '\n')]

- Note that all the base forms extracted with spaCy are accurate and carry usable meanings.

5.5.1.2 The spaCy Lemmatizer

The spaCy lemmatizer is known for its high speed and efficiency. It is optimized to swiftly process large amounts of text data. NLTK also stands for a solid performance but tends to be slower as compared to spaCy when processing huge of text data. The following is a code demo.

```
import spacy

# These English pipelines have an inbuilt rule-based lemmatizer.
nlp = spacy.load("en_core_web_sm")
lemmatizer = nlp.get_pipe("lemmatizer")
print(lemmatizer.mode)   # 'rule'

sample_doc = nlp(""" The SpaCy lemmatizer is designed using a combination of rule-based and statistical methods to find the
                    base forms of words. It comes integrated
                    with the SpaCy NLP library.""")
print([token.lemma_ for token in sample_doc])
rule
[' ', 'the', 'SpaCy', 'lemmatizer', 'be', 'design', 'use', 'a', 'combination', 'of', 'rule', '-', 'base', 'and', 'statistical', 'method', '\n                    ', 'to', 'find', 'the', 'base', 'form', 'of', 'word', '.', 'it', 'come', 'integrate', 'with', 'the', 'SpaCy', 'NLP', 'library', '.']
```

5.5.1.3 TextBlob Lemmatizer

TextBlob performs faster when compared to NLTK. It can be easily deployed with fewer computing resources. TextBlob is simpler to use, and it supports many functions that are not available in NLTK. Listing 5-3 is its code demo.

Listing 5-3. Program to Demonstrate TextBlob Lemmatizer

```
# First install TextBlob.
!pip install textblob
from textblob import TextBlob, Word

# Create sample text.
text = """ TextBlob performs faster when compared to nltk. It
can be easily deployed with lesser computing resources.
        TextBlob is simpler to use and it supports many
functions that are not available in nltk.
    """

# Create a TextBlob object.
blob = TextBlob(text)

# Tokenize the sample text.
words = blob.words

# Lemmatize.
lemmatized_words = [Word(word).lemmatize() for word in words]

# Print the output.
print("Original sample words:", words)
print("Lemmatized words:", lemmatized_words)
OUTPUT:
Original sample words: ['TextBlob', 'performs', 'faster',
'when', 'compared', 'to', 'nltk', 'It', 'can', 'be', 'easily',
```

'deployed', 'with', 'lesser', 'computing', 'resources',
'TextBlob', 'is', 'simpler', 'to', 'use', 'and', 'it',
'supports', 'many', 'functions', 'that', 'are', 'not',
'available', 'in', 'nltk']
Lemmatized words: ['TextBlob', 'performs', 'faster', 'when',
'compared', 'to', 'nltk', 'It', 'can', 'be', 'easily',
'deployed', 'with', 'lesser', 'computing', 'resource',
'TextBlob', 'is', 'simpler', 'to', 'use', 'and', 'it',
'support', 'many', 'function', 'that', 'are', 'not',
'available', 'in', 'nltk']

5.6 A Complete Lexical Analysis with Data Preprocessing

You have already learned enough text preprocessing steps to tackle a complete text analysis problem. Next, we cover text preprocessing and predictive analytics. It's a rather long problem.

As this is a long problem, please explain the steps you are taking and why something has to be done.

5.6.1 YouTube Comments Spam Detection

We solved the same problem in the Chapter 1 code demo. This time, we attempt to solve it here again with some data preprocessing steps and see if we get any improvements in results.

```
# Import the required libraries.
import pandas as pd
import numpy as np
# The below code is for working with machine learning model.
from sklearn import feature_extraction, linear_model, model_
```

```
selection, preprocessing
from sklearn.model_selection import train_test_split
from sklearn.ensemble import RandomForestClassifier
from sklearn.metrics import accuracy_score
# Ignore warnings.
import warnings
warnings.filterwarnings('ignore')
import re
import nltk
import spacy
import string
pd.options.mode.chained_assignment = None
# The below code is for working with machine learning model.
from sklearn import feature_extraction, linear_model, model_
selection, preprocessing
from sklearn.model_selection import train_test_split
from sklearn.ensemble import RandomForestClassifier
from sklearn.metrics import accuracy_score
```

Let's take a quick look at our data.

```
# Read the data files [1] available in the same folder as
this code.
Youtube01_psy = pd.read_csv('Youtube01-Psy.csv')
Youtube02_katyperry = pd.read_csv('Youtube02-KatyPerry.csv')
Youtube03_lmfao = pd.read_csv('Youtube03-LMFAO.csv')
Youtube04_eminem = pd.read_csv('Youtube04-Eminem.csv')
Youtube05_shakira = pd.read_csv('Youtube05-Shakira.csv')
# ACombine all five datasets.
combined_df = pd.concat([Youtube01_psy, Youtube02_katyperry,
Youtube03_lmfao, Youtube04_eminem, Youtube05_shakira])
```

CHAPTER 5   LEXICAL ANALYSIS

```
# Reset the index
combined_df.reset_index(drop=True, inplace=True)
combined_df.head(3)
                                     COMMENT_ID            AUTHOR  \
0  LZQPQhLyRh8OUYxNuaDWhIGQYNQ96IuCg-AYWqNPjpU        Julius NM
1  LZQPQhLyRh_C2cTtd9MvFRJedxydaVW-2sNg5Diuo4A       adam riyati
2  LZQPQhLyRh9MSZYnf8djyk0gEF9BHDPYrrK-qCczIY8   Evgeny Murashkin

                  DATE  \
      CONTENT  \
0  2013-11-07T06:20:48  Huh, anyway check out this you[tube]
                        channel: ...
1  2013-11-07T12:37:15  Hey guys check out my new channel and
                        our firs...
2  2013-11-08T17:34:21  just for test I have to say murdev.com

   CLASS
0      1
1      1
2      1
# Keep only the useful "CONTENT" and "CLASS" columns.
combined_df = combined_df[["CONTENT", "CLASS"]]
# Randomly select 5 rows
random_sample = combined_df.sample(n=5)
print(random_sample)
                                             CONTENT  CLASS
1127  The best Song i saw ♥♥♥♥♥♥♥♥☺☺☺☺☺☺☺☺☺...      0
599   Hey Katycats! We are releasing a movie at midn...      1
574   want to win borderlands the pre-sequel? check ...      1
533   Awesome video this is one of my favorite  song...      0
747           Love this video and the song of course       0
```

CHAPTER 5 LEXICAL ANALYSIS

- Note the special characters and misalignment due to spaces in CLASS

```python
# Randomly select 5 rows
random_sample = combined_df.sample(n=5)
print(random_sample)
```
```
                                               CONTENT  CLASS
532    http://www.googleadservices.com/pagead/aclk?sa...      1
1447             I love this song sooooooooooooooo much      0
1901   Hey youtubers... I really appreciate all of yo...      1
268    https://www.facebook.com/pages/Mathster-WP/149...      1
1304   sorry but eminmem is a worthless wife beating ...      0
```

5.6.2 Convert the Text to Lowercase

```python
# Convert all comments to string type for further processing.
combined_df["CONTENT"] = combined_df["CONTENT"].astype(str)
# Convert everything in to lower case.
combined_df["text_lower_case"] = combined_df["CONTENT"].str.lower()
# Randomly select 3 rows.
random_sample_lower = combined_df.sample(n=3)
print(random_sample_lower)
```
```
                                     CONTENT  CLASS \
404   YAY IM THE 11TH COMMENTER!!!!!             ...      1
907               Check out this playlist on YouTube:      1
942   View 851.247.920<br /><br /> Best youtube Vide...      1

                              text_lower_case
404   yay im the 11th commenter!!!!!        ...
```

```
907                    check out this playlist on youtube:
942   view 851.247.920<br /><br /> best youtube vide...
# Drop ombined_df["CONTENT"] as we will work only with
combined_df["text_lower_case"].
# The punctuations present are - !"#$%&\'()*+,-./:;<=>?@
[\\]^_{|}~`
# combined_df = combined_df.drop(columns=["CONTENT"])
combined_df.columns
Index(['CONTENT', 'CLASS', 'text_lower_case'], dtype='object')
```

5.6.3 Remove All Unwanted Punctuation

```
# Punctuation to remove
punctuation_to_remove = "!\"#$%&\'()*+,-./:;<=>?@[\\]^_{|}~`"
```

```
# Remove the above punctuations from the # Remove the above
punctuations from the "text_lower_case" column.
combined_df["text_lower_case"] = combined_df["text_lower_
case"].str.translate(str.maketrans('', '', punctuation_to_
remove))
# Randomly select 10 rows.
random_sample_lower = combined_df.sample(n=10)
print(random_sample_lower)
                                          CONTENT CLASS \
173               http://www.gofundme.com/gvr7xg        1
277    Hey, join me on tsū, a publishing platform whe...  1
1828                    Shakira is very beautiful        0
25     marketglory . com/strategygame/andrijamatf ear... 1
580    Thank you KatyPerryVevo for your instagram lik... 1
1240   all u should go check out j rants vi about eminem 1
1788              Please visit this Website: oldchat.tk  1
962    <br />Please help me get 100 subscribers by t... 1
```

1857	Love it	0
301	http://hackfbaccountlive.com/?ref=4436607 psy...	1

	text_lower_case
173	httpwwwgofundmecomgvr7xg
277	hey join me on tsū a publishing platform where...
1828	shakira is very beautiful
25	marketglory comstrategygameandrijamatf earn r...
580	thank you katyperryvevo for your instagram lik...
1240	all u should go check out j rants vi about eminem
1788	please visit this website oldchattk
962	br please help me get 100 subscribers by the ...
1857	love it
301	httphackfbaccountlivecomref4436607 psy news o...

5.6.4 Remove Stop Words

```
import nltk
from nltk.corpus import stopwords

# Download the stopwords if you haven't already
nltk.download('stopwords')

# Get the list of English stopwords
stop_words = set(stopwords.words('english'))
```
[nltk_data] Downloading package stopwords to C:\Users\Shailendra
[nltk_data]   Kadre\AppData\Roaming\nltk_data...
[nltk_data]   Package stopwords is already up-to-date!
```
combined_df.columns
```
Index(['CONTENT', 'CLASS', 'text_lower_case'], dtype='object')
```
# Remove stopwords from the "text_lower_case" column
```

CHAPTER 5 LEXICAL ANALYSIS

```
combined_df["text_lower_case"] = combined_df["text_lower_
case"].apply(lambda x: ' '.join([word for word in x.split() if
word not in stop_words]))
# Randomly select 5 rows.
random_sample_lower = combined_df.sample(n=5)
print(random_sample_lower)
```

```
                                               CONTENT  CLASS \
1025   <a href="https://m.freemyapps.com/share/url/10...        1
181                          Please check out my vidios        1
724    This awesome song needed 4 years to reach to 8...        0
1296   5 years and i still dont get the music video h...        0
878                                                 omg        0

                                                text_lower_case
1025   hrefhttpsmfreemyappscomshareurl10b35481httpsmf...
181                                   please check vidios
724      awesome song needed 4 years reach 800 mil view...
1296       5 years still dont get music video help someone
878                                                    omg
```

The frequent or rare word removal is based on the specific goals of your NLP task and the nature of the dataset in hand. Frequent words are removed because of their minimal informational value. Removing frequent and rare words helps to reduce noise in the data and allows focus on more meaningful words. It improves the performance of several NLP algorithms as they can now focus on content-rich words that contribute to the overall analysis. Rare words are particularly removed as they have limited relevance.

Next, let's look at code on how to remove frequent and rare words.

5.6.5 Remove Frequent Words

```python
from collections import Counter

# Combine all text into a single string and split into
individual words.
all_words = ' '.join(combined_df["text_lower_case"]).split()

# Count the frequency of every word.
word_counts = Counter(all_words)

# Determine the threshold for frequent words (top 10 most
common words).
most_common_words = word_counts.most_common(10)

# Print the frequent words with their frequencies.
print("Frequent words with their frequencies:")
for word, count in most_common_words:
    print(f"{word}: {count}")

# Set of the frequent words.
frequent_words = {word for word, count in most_common_words}

# Remove frequent words from the combined_df["text_
lower_case"].
combined_df["frequent_removed"] = combined_df["text_lower_
case"].apply(
    lambda x: ' '.join([word for word in x.split() if word not
    in frequent_words])
)
```
Frequent words with their frequencies:
check: 559
video: 294
: 267
like: 235

```
please: 231
song: 231
subscribe: 209
love: 189
channel: 173
music: 144
combined_df.columns
Index(['CONTENT', 'CLASS', 'text_lower_case', 'frequent_
removed'], dtype='object')
# Randomly select 5 rows.
random_sample_lower = combined_df[['CLASS', 'frequent_
removed']].sample(n=5)
print(random_sample_lower)
        CLASS                            frequent_removed
1939        1  peoples earth seen perform every form evil lei...
459         0  comment randomly get lots likes replies reason...
896         0                    almost 1 billion views nice
1799        0                                   she39s pretty
1426        0              charlieee dddd saw lost understand
```

5.6.6 Remove Rare Words

```
from collections import Counter
# Combine all text into a single string and split into
individual words.
all_words = ' '.join(combined_df["text_lower_case"]).split()

# Count the frequency of every word.
word_counts = Counter(all_words)

# Define the threshold for rare words.
threshold = 5
```

```
rare_words = {word: count for word, count in word_counts.
items() if count < threshold}
```

```
# Sort the rare words by frequency in ascending order and keep
the top 5.
sorted_rare_words = sorted(rare_words.items(), key=lambda item:
item[1])
top_5_rare_words = sorted_rare_words[:5]
```

```
# Print the top 5 rare words with their frequencies.
print("Top 5 rare words with their frequencies:")
for word, count in top_5_rare_words:
    print(f"{word}: {count}")
```

```
# Remove rare words from combined_df["text_lower_case"].
combined_df["rare_removed"] = combined_df["text_lower_
case"].apply(
    lambda x: ' '.join([word for word in x.split() if word not
    in rare_words])
)
Top 5 rare words with their frequencies:
anyway: 1
kobyoshi02: 1
monkeys: 1
shirtplease: 1
test: 1
```

- All the removed rare words are not displayed here as the list is very long.

    ```
    combined_df.columns
    ```

    ```
    Index(['CLASS', 'text_lower_case', 'frequent_removed',
    'rare_removed'], dtype='object')
    ```

CHAPTER 5 LEXICAL ANALYSIS

```
# Randomly select 5 rows.
random_sample_lower = combined_df[['CLASS', 'lower_
removed']].sample(n=5)
print(random_sample_lower)
      CLASS                                   lower_removed
1455  0                                                  so
1787  1                         please visit this website
35    0     why is a korean song so big in the does that m...
1622  1                     check out this video on youtube
1386  0
```

Next, let's apply stemming and lemmatization directly to combined_df["text_lower_case"] because they are necessary steps.

5.6.7 Stemming

```
import nltk
from nltk.stem import PorterStemmer
from collections import Counter

# Ensure that the necessary NLTK resources are available
nltk.download('punkt')

# Initialize the Porter Stemmer
stemmer = PorterStemmer()

# Function to stem words in a text and return the
stemmed version
def stem_text(text):
    words = text.split()
    stemmed_words = [stemmer.stem(word) for word in words]
    return ' '.join(stemmed_words)
[nltk_data] Downloading package punkt to C:\Users\Shailendra
```

```
[nltk_data]        Kadre\AppData\Roaming\nltk_data...
[nltk_data]     Package punkt is already up-to-date!
# Apply stemming to tcombined_df["text_lower_case"].
combined_df["text_lower_case"] = combined_df["text_lower_
case"].apply(stem_text)
# Combine all stemmed text into a single string and split into
individual words
all_stemmed_words = ' '.join(combined_df["text_lower_
case"]).split()

# Count the frequency of every stemmed word
word_counts = Counter(all_stemmed_words)

# Get the top 5 stemmed words and their frequencies
top_5_stemmed_words = word_counts.most_common(5)
# Create a mapping from stemmed words to their original
base forms
base_form_mapping = {}
for text in combined_df["text_lower_case"]:
    words = text.split()
    for word in words:
        stemmed_word = stemmer.stem(word)
        if stemmed_word not in base_form_mapping:
            base_form_mapping[stemmed_word] = set()
        base_form_mapping[stemmed_word].add(word)
# Display the top 5 stemmed words with their original
base forms
print("Top 5 stemmed words with their original base forms:")
for stemmed_word, count in top_5_stemmed_words:
    base_forms = base_form_mapping.get(stemmed_word, [])
    base_form_display = ', '.join(base_forms)  # Display unique
base forms
```

CHAPTER 5 LEXICAL ANALYSIS

```
    print(f"Stemmed Word: {stemmed_word}, Count: {count},
Original Words: {base_form_display}")
Top 5 stemmed words with their original base forms:
Stemmed Word: check, Count: 568, Original Words: check
Stemmed Word: video, Count: 361, Original Words: video
Stemmed Word: song, Count: 274, Original Words: song
Stemmed Word: , Count: 267, Original Words:
Stemmed Word: like, Count: 256, Original Words: like
combined_df.columns
Index(['CONTENT', 'CLASS', 'text_lower_case', 'frequent_
removed',
       'rare_removed'],
      dtype='object')
combined_df[['CLASS', 'text_lower_case']].head(10)
   CLASS                                    text_lower_case
0      1          huh anyway check youtub channel kobyoshi02
1      1      hey guy check new channel first vid us monkey ...
2      1                                  test say murdevcom
3      1                          shake sexi ass channel enjoy
4      1                              watchvvtarggvgtwq check
5      1     hey check new websit site kid stuff kidsmediau...
6      1                                    subscrib channel
7      0              turn mute soon came want check views
8      1                          check channel funni videos
9      1                   u shouldd check channel tell next
# Randomly select 5 stemmed rows.
random_sample = combined_df[['CLASS', 'text_lower_case']].
sample(n=5)
print(random_sample)
        CLASS                                   text_lower_case
1749       1          brazil pleas subscrib channel love all
```

602	0	song never get old lt3
1329	1	guy check extraordinari websit call zonepacom ...
1791	1	hello guysi found way make money onlin get pai...
1479	0	anybodi els 2015

5.6.8 Lemmatization

We can perform lemmatization in two ways.

- Without POS tagging: It is less accurate as this way, the lemmatizer often assumes that words are nouns, which can lead to potential errors.

- With POS tagging: It is more accurate as it takes the word's role in the sentence into account while performing lemmatization. Taking POS tags in to account reduces the likelihood of errors.

The following code focuses on the second process. A note on the coding approach: POS tags from NLTK are more detailed compared to the broader ones from WordNet. NLTK's POS tagger provides the necessary contextual information. Converting NLTK's detailed tags to WordNet's simpler POS tags ensures that the lemmatizer has the necessary context.

```
import nltk
from nltk.corpus import wordnet
from nltk.stem import WordNetLemmatizer
from nltk import pos_tag, word_tokenize

# Convert NLTK POS tags to WordNet POS tags as explained above.
def get_wordnet_pos(treebank_tag):
    if treebank_tag.startswith('J'):
        return wordnet.ADJ
```

CHAPTER 5 LEXICAL ANALYSIS

```python
    elif treebank_tag.startswith('V'):
        return wordnet.VERB
    elif treebank_tag.startswith('N'):
        return wordnet.NOUN
    elif treebank_tag.startswith('R'):
        return wordnet.ADV
    else:
        return wordnet.NOUN  # Assume nounun if no match is found.
# Test on sample text.
text = """Note on the approach of coding: POS tags from NLTK's are more detailed
            compared to the broader ones from WordNet. NLTK's POS
            tagger provides the necessary contextual information.
            Converting NLTK's detailed tags to WordNet' simpler s
            POS tags ensures that the lemmatizer has the
            necessary context.
"""

# Tokenize and find the POS tags.
tokens = word_tokenize(text)
pos_tags = pos_tag(tokens)

# Initialize the lemmatizer.
lemmatizer = WordNetLemmatizer()

# Lemmatize with POS tags
lemmatized_words = [lemmatizer.lemmatize(token, get_wordnet_pos(pos)) for token, pos in pos_tags]

print(lemmatized_words)
['Note', 'on', 'the', 'approach', 'of', 'coding', ':', 'POS',
'tag', 'from', 'NLTK', "'", 's', 'be', 'more', 'detailed',
```

'compare', 'to', 'the', 'broad', 'one', 'from', 'WordNet', '.', 'NLTK', ''', 's', 'POS', 'tagger', 'provide', 'the', 'necessary', 'contextual', 'information', '.', 'Converting', 'NLTK', ''', 's', 'detail', 'tag', 'to', 'WordNet', "'s", 'simpler', 'POS', 'tag', 'ensure', 'that', 'the', 'lemmatizer', 'have', 'the', 'necessary', 'context', '.']
combined_df.columns
Index(['CONTENT', 'CLASS', 'text_lower_case', 'frequent_removed', 'rare_removed'], dtype='object')

```
# We will aplly the lemmatize diectly on combined_df['text_lower_case'] as its a necessary step in our case.
# First write a function to lemmatize text.
def lemmatize_text(text):
    tokens = word_tokenize(text)
    pos_tags = pos_tag(tokens)
    lemmatized_tokens = [lemmatizer.lemmatize(token, get_wordnet_pos(pos)) for token, pos in pos_tags]
    return ' '.join(lemmatized_tokens)
# Apply the lemmatization function directly to combined_df['text_lower_case'].
combined_df['text_lower_case'] = combined_df['text_lower_case'].apply(lemmatize_text)

# Display the DataFrame with the new lemmatized column
print(combined_df[['CLASS','text_lower_case']])
      CLASS                                    text_lower_case
0         1        huh anyway check youtub channel kobyoshi02
1         1     hey guy check new channel first vid u monkey i...
2         1                                    test say murdevcom
3         1                         shake sexi as channel enjoy
```

CHAPTER 5 LEXICAL ANALYSIS

```
4         1                    watchvvtarggvgtwq check
...       ...                                      ...
1951      0                     love song sing camp time
1952      0   love song two reason 1it africa 2i born beauti...
1953      0                                              wow
1954      0                              shakira u wiredo
1955      0                            shakira best dancer

[1956 rows x 2 columns]
```

We are still left with the following text preprocessing processes that we directly apply to our main text column, combined_df['text_lower_case'].

- Removal of emojis
- Removal of emoticons
- Conversion of emoticons to words
- Conversion of emojis to words
- Removal of URLs
- Removal of HTML tags
- Chat words conversion
- Spelling correction

Let's examine each.

5.6.9 Removal of Emojis

In text preprocessing, we need to remove emojis to simplify text data. Removing emojis lets models focus on the most relevant content. Removing emojis helps us to standardize the input data and makes it more uniform and easier to process.

 # We will use regular expressions to remove emojis directly from combined_df['text_lower_case'].

```python
import pandas as pd
import re

# Function to remove emojis
def remove_emojis(text):
    emoji_pattern = re.compile(
        "["
        u"\U0001F600-\U0001F64F"  # Emoticons
        u"\U0001F300-\U0001F5FF"  # Symbols & Pictographs
        u"\U0001F680-\U0001F6FF"  # Transport & Map Symbols
        u"\U0001F1E0-\U0001F1FF"  # Flags (iOS)
        u"\U00002702-\U000027B0"  # Miscellaneous Symbols
        u"\U000024C2-\U0001F251"  # Enclosed Characters
        "]+", flags=re.UNICODE)
    return emoji_pattern.sub(r'', text)
# Apply the function diectly to combined_df['text_lower_case'].
combined_df['emojis_removed'] = combined_df['text_lower_case'].apply(remove_emojis)

# Pring the output.
print(combined_df[['CLASS','emojis_removed']])
        CLASS                              emojis_removed
0           1        huh anyway check youtub channel kobyoshi02
1           1        hey guy check new channel first vid u monkey i...
2           1                                  test say murdevcom
3           1                      shake sexi as channel enjoy
4           1                          watchvvtarggvgtwq check
...        ...                                             ...
1951        0                          love song sing camp time
1952        0        love song two reason 1it africa 2i born beauti...
```

CHAPTER 5   LEXICAL ANALYSIS

```
1953          0                                           wow
1954          0                              shakira u wiredo
1955          0                           shakira best dancer

[1956 rows x 2 columns]
combined_df.columns
Index(['CONTENT', 'CLASS', 'text_lower_case', 'frequent_removed',
       'rare_removed', 'emojis_removed'], dtype='object')
```

5.6.10 Removal of Emoticons

You might be wondering the difference between emojis and emoticons. Emojis are colorful, digital icons like ☺ (smiling face) or 🚀 (rocket). They stand for objects or emotions. While emoticons are text-based symbols like :-) (smiley face) or <3 (heart). Emoticons are created with keyboard characters to express feelings.

```python
import re

# Apply RegEx to remove emoticons. Write a function first.
def remove_emoticons(text):
    # RegEx pattern to match commonly used emoticons.
    emoticon_pattern = re.compile(
        r'[:;=8][\-o\*]?[)\]\(\[dDpP\|/\\\^]',
        re.UNICODE)
    return emoticon_pattern.sub(r'', text)
# Sample sentence with emoticons.
sample_sentence = "Hello John! :) How are you? :P I hope you're good today. :D"
```

CHAPTER 5 LEXICAL ANALYSIS

```
# Remove emoticons from the sample sentence
emoticons_removed = remove_emoticons(sample_sentence)

# Print the result
print("Original Sentence:", sample_sentence)
print("Cleaned Sentence:", emoticons_removed)
```
Original Sentence: Hello John! :) How are you? :P I hope you're good today. :D
Cleaned Sentence: Hello John! How are you? I hope you're good today.
```
# Apply the function to the combined_df['text_lower_case']. We will store it in another column for now.
combined_df['emoticons_removed'] = combined_df['text_lower_case'].apply(remove_emoticons)

# Print the output.
print(combined_df[['CLASS','emoticons_removed']])
```
```
         CLASS                          emoticons_removed
0            1         huh anyway check youtub channel kobyoshi02
1            1         hey guy check new channel first vid u monkey i...
2            1                                    test say murdevcom
3            1                            shake sexi as channel enjoy
4            1                               watchvvtarggvgtwq check
...        ...                                                   ...
1951         0                              love song sing camp time
1952         0         love song two reason 1it africa 2i born beauti...
1953         0                                                    wow
1954         0                                       shakira u wiredo
1955         0                                    shakira best dancer

[1956 rows x 2 columns]
```

195

```
combined_df.columns
Index(['CONTENT', 'CLASS', 'text_lower_case', 'frequent_removed',
       'rare_removed', 'emojis_removed', 'emoticons_removed'],
      dtype='object')
```

5.6.10.1 Converting Emoticons to Words

We can map emoticons to their corresponding descriptions, like "smiley face" or "grinning face." This conversion helps preserve the sentiments conveyed by emoticons in the form of simple text words, which are easier to analyze compared to plain emoticons. It helps improve the accuracy and effectiveness of NLP tasks.

```
# Map the common emoticons to words.
emoticon_to_word = {
    ':)': 'smiley face',
    ':D': 'grinning face',
    ':P': 'playful face',
    ':-)': 'smiley face',
    ':-D': 'grinning face',
    ':-P': 'playful face'
}
# Function to replace emoticons with words.
def emoticons_to_word_converter(text):
    for emoticon, word in emoticon_to_word.items():
        text = text.replace(emoticon, word)
    return text
# Sample sentence.
sample_sentence = "Hello there! :) How are you? :P I hope you're doing well. :D"
```

```
# Convert emoticons to words.
converted_sentence = emoticons_to_word_converter(sample_sentence)

# Print the output.
print("Original Sentence:", sample_sentence)
print("Converted Sentence:", converted_sentence)
Original Sentence: Hello there! :) How are you? :P I hope
you're doing well. :D
Converted Sentence: Hello there! smiley face How are you?
playful face I hope you're doing well. grinning face
# Apply the function directly on combined_df['text_lower_case']
# Its likely to be useful in increasing the accuracy of our
analysis.

# Apply the function to the 'text_lower_case' column
combined_df['text_lower_case'] = combined_df['text_lower_
case'].apply(convert_emoticons_to_words)

# Print the output.
print(combined_df[['CLASS', 'text_lower_case']])
      CLASS                               text_lower_case
0         1        huh anyway check youtub channel kobyoshi02
1         1     hey guy check new channel first vid u monkey i...
2         1                                   test say murdevcom
3         1                       shake sexi as channel enjoy
4         1                             watchvvtarggvgtwq check
...     ...                                               ...
1951      0                         love song sing camp time
1952      0   love song two reason 1it africa 2i born beauti...
1953      0                                               wow
1954      0                                   shakira u wiredo
1955      0                                shakira best dancer

[1956 rows x 2 columns]
```

5.6.10.2 Conversion of Emojis to Words

```
# Install emoji if you have not done it earlier.
!pip install emoji # it is the necessarylibrary.
Collecting emoji
  Downloading emoji-2.12.1-py3-none-any.whl (431 kB)
                                 0.0/431.4 kB ? eta -:--:--
     -------                     81.9/431.4 kB 2.3 MB/s eta 0:00:01
     ---------------------       256.0/431.4 kB 3.2 MB/s eta 0:00:01
     ---------------------       431.4/431.4 kB 3.4 MB/s eta 0:00:00
Requirement already satisfied: typing-extensions>=4.7.0 in c:\
users\shailendra kadre\anaconda3\lib\site-packages (from emoji)
(4.12.2)
Installing collected packages: emoji
Successfully installed emoji-2.12.1
import emoji

# Map of emojis to words.
def emojis_to_word_converer(text):
    # Convert emojis to their corresponding descriptions.
    return emoji.demojize(text)
# Example usage
sample_text = "Hello John! ☺ How are you? 🚀 I hope you're good today. 🎉"
converted_text = emojis_to_word_converer(sample_text)

print("Original Text:", sample_text)
print("Converted Text:", converted_text)
Original Text: Hello John! ☺ How are you? 🚀 I hope you're good today. 🎉
Converted Text: Hello John! :smiling_face_with_smiling_eyes: How are you? :rocket: I hope you're good today. :party_popper:
# Apply the function to combined_df['text_lower_case'].
```

```
combined_df['text_lower_case'] = combined_df['text_lower_
case'].apply(emojis_to_word_converer)
```

```
# Print the output.
print(combined_df[['CLASS', 'text_lower_case']])
      CLASS                                  text_lower_case
0         1       huh anyway check youtub channel kobyoshi02
1         1   hey guy check new channel first vid u monkey i...
2         1                                test say murdevcom
3         1                     shake sexi as channel enjoy
4         1                       watchvvtarggvgtwq check
...     ...                                              ...
1951      0                       love song sing camp time
1952      0  love song two reason 1it africa 2i born beauti...
1953      0                                              wow
1954      0                               shakira u wiredo
1955      0                             shakira best dancer

[1956 rows x 2 columns]
```

5.6.11 Removal of URLs

URLs are noise and irrelevant information for any text analysis. Removing URLs standardizes and cleans the input data.

```
import re

# Let's write a function to remove URLs from the input text.
def url_remover(text):
    url_pattern = re.compile(r'http[s]?://\S+|www\.\S+')
    return url_pattern.sub('', text)
# Sample usage below.
sample_text = "Check out this link: https://www.example.com and
also visit http://example.org."
```

```
cleaned_text = url_remover(sample_text)
```

```
print("Original Text:", sample_text)
print("Cleaned Text:", cleaned_text)
Original Text: Check out this link: https://www.example.com and
also visit http://example.org.
Cleaned Text: Check out this link:  and also visit
# Apply the function to combined_df['text_lower_case'].
combined_df['text_lower_case'] = combined_df['text_lower_
case'].apply(url_remover)
```

```
# Print the output.
print(combined_df[['CLASS','text_lower_case']])
      CLASS                                    text_lower_case
0         1          huh anyway check youtub channel kobyoshi02
1         1       hey guy check new channel first vid u monkey i...
2         1                                     test say murdevcom
3         1                         shake sexi as channel enjoy
4         1                             watchvvtarggvgtwq check
...     ...                                                 ...
1951      0                             love song sing camp time
1952      0  love song two reason 1it africa 2i born beauti...
1953      0                                                  wow
1954      0                                     shakira u wiredo
1955      0                                  shakira best dancer

[1956 rows x 2 columns]
```

5.6.12 Removal of HTML tags

HTML tags may pose as noise and irrelevant data in most text analysis. Removing them can improve the accuracy of our text analysis.

```python
import re

# A simple function to remove HTML tags from the input text.
def html_tag_remover(text):
    html_tag_pattern = re.compile(r'<[^>]+>')
    return html_tag_pattern.sub('', text)
# Sample usage.
sample_text = "<p>Hi John! <a href='https://example_url.com'>Click here</a> to visit.</p>"
cleaned_text = html_tag_remover(sample_text)

print("Original Text:", sample_text)
print("Cleaned Text:", cleaned_text)
Original Text: <p>Hi John! <a href='https://example_url.com'>Click here</a> to visit.</p>
Cleaned Text: Hi John! Click here to visit.
# Apply the function to combined_df['text_lower_case'].
combined_df['text_lower_case'] = combined_df['text_lower_case'].apply(html_tag_remover)

# Print the output.
print(combined_df[['CLASS','text_lower_case']])
```

	CLASS	text_lower_case
0	1	huh anyway check youtub channel kobyoshi02
1	1	hey guy check new channel first vid u monkey i...
2	1	test say murdevcom
3	1	shake sexi as channel enjoy
4	1	watchvvtarggvgtwq check
...	...	...
1951	0	love song sing camp time
1952	0	love song two reason 1it africa 2i born beauti...
1953	0	wow

CHAPTER 5   LEXICAL ANALYSIS

```
1954        0                          shakira u wiredo
1955        0                        shakira best dancer

[1956 rows x 2 columns]
```

5.6.13 Chat Words Conversion

Chat words are informal, abbreviated, or slang terms that are popularly used in online messaging, especially by the younger generation. The examples include "u" for "you" or "lol" for "laughing out loud." Converting chat words into more formal language words can help in accurately analyzing them by text processing models.

```python
# Map common chat words to more formal words. You can add on to this list.
chat_to_formal = {
    'u': 'you',
    'r': 'are',
    'lol': 'laughing out loud',
    'brb': 'be right back',
    'ttyl': 'talk to you later',
    'thx': 'thanks',
    'gtg': 'got to go',
    'b4': 'before'
}

# Let's now write a function to replace chat words with formal words.
def chat_word_converter(text):
    for chat_word, formal_word in chat_to_formal.items():
        text = text.replace(chat_word, formal_word)
    return text
```

```
# Sample usage.
sample_text = "Hey John! r u coming to the function this
evening? lol, thx for your invite!"
converted_text = chat_word_converter(sample_text)

print("Original Text:", sample_text)
print("Converted Text:", converted_text)
Original Text: Hey John! r u coming to the function this
evening? lol, thx for your invite!
Converted Text: Hey John! are you coming to the fyounction this
evening? laughing out loud, thanks foare yoyouare invite!
# Apply the function to combined_df['text_lower_case'].
combined_df['text_lower_case'] = combined_df['text_lower_
case'].apply(chat_word_converter)

# Print the output.
print(combined_df[['CLASS','text_lower_case']])
      CLASS                               text_lower_case
0         1  hyouh anyway check yoyoutyoub channel kobyoshi02
1         1  hey gyouy check new channel fiarest vid you mo...
2         1                            test say myouaredevcom
3         1                      shake sexi as channel enjoy
4         1                      watchvvtaareggvgtwq check
...     ...                                            ...
1951      0                        love song sing camp time
1952      0  love song two areeason 1it afareica 2i boaren ...
1953      0                                             wow
1954      0                           shakiarea you wiareedo
1955      0                         shakiarea best danceare

[1956 rows x 2 columns]
```

5.6.14 Spelling Corrections

Spelling correction increases information gain and helps with accurate analysis of the input text.

```
!pip install SpellChecker # Necesary library.
!pip install indexer # Another necesary library.
!pip install pyspellchecker # Another necesary library.
from spellchecker import SpellChecker
import pandas as pd

# Initialize the spell checker
spell = SpellChecker()

# Function to correct spelling in text
def spelling_corrector(text):
    words = text.split()  # Split the text into words
    corrected_words = [spell.candidates(word).pop() if spell.candidates(word) else word for word in words]
    return ' '.join(corrected_words)
# Sample usage.
sample_text = """We havv few speling mistkes in this short paragraph.',
                    'Anothr exampl with spellig erors.',
                    'No spellig errors here!
                """
converted_text = spelling_corrector(sample_text)

print("Original Text:", sample_text)
print("Converted Text:", converted_text)
Original Text: I havv a speling mistke in this sentnce.',
                    'Anothr exampl with spellig
                    erors.',
                    'No spellig errors here!
```

Converted Text: I have a spieling mistake in this sentnce.',
another example with spelling erors.', no spelling errors heres
 # Apply the function to combined_df['text_lower_case'].
combined_df['text_lower_case'] = combined_df['text_lower_case'].apply(spelling_corrector)

 # Print the output.
print(combined_df[['CLASS','text_lower_case']])

Note Applying the spelling_corrector function to the combined_df['text_lower_case'] was taking a long time. So, I aborted the kernel. You can try it on a more powerful machine or in the cloud. We also tried TextBlob, but it took a lot of time.

---

combined_df.columns
Index(['CONTENT', 'CLASS', 'text_lower_case', 'frequent_removed',
       'rare_removed', 'emojis_removed', 'emoticons_removed'],
      dtype='object')

- The useful columns for further processing are 'CLASS' and 'text_lower_case'. All others are for demo. You can use their code in your projects depending on what type of text analysis you are trying to do.

5.6.15 Apply Machine Learning Model

Listing 5-4 is similar to what you saw in Chapter 1.

Lisitng 5-4. Code Demonstration of Applying Machine Learning Model

```
# Seperate features and the target.
X = np.array(combined_df['text_lower_case'])
y = np.array(combined_df['CLASS'])
count_vectorizer = feature_extraction.text.CountVectorizer()
# Instrantiate CountVectorizer()
X.shape
(1956,)
# Split in to train and test datasets.
X_train, X_test, y_train, y_test = train_test_split(X, y,
test_size=0.3, random_state=42)
# We will use count_vectorizer.fit_transform() for X_train
and X_test.
X_train = count_vectorizer.fit_transform(X_train)
X_test = count_vectorizer.transform(X_test) # We will do onlt
transform() with X_test.
# Initialize the Random Forest model
clf = RandomForestClassifier(n_estimators=100, random_state=42)
scores = model_selection.cross_val_score(clf, X_train, y_train,
cv=3, scoring="f1")
scores
array([0.87470449, 0.9044289 , 0.90487239])
```

CHAPTER 5　LEXICAL ANALYSIS

> **Note** Surprisingly, we got considerably better accuracies in Chapter 1, in which minimal preprocessing was done. Looks like there is a considerable loss of information in these preprocessing steps. It's an interesting lesson for all of us. With text preprocessing, the results are expected to improve in the normal course. It can happen only if the right and only the necessary steps are applied while preprocessing. Overly preprocessing data can cause it to lose the information, and it can lead to poor results.

```
clf.fit(X_train, y_train) # Fit the model.
RandomForestClassifier(random_state=42)
# Make predictions with the test data.
y_pred = clf.predict(X_test)
# Construct a dataframe with columns as y_test and y_pred.
test_df = pd.DataFrame()
test_df["y"] = y_test
test_df["y_predict"] = y_pred
# Display 10 random rows from test_df
random_sample = test_df.sample(n=10)
print(random_sample)
     y  y_predict
9    1  1
218  1  1
3    1  1
362  1  1
291  1  1
174  0  0
99   1  1
```

CHAPTER 5   LEXICAL ANALYSIS

```
357  0           0
531  0           0
294  1           1
```

- The ground truth, y_test, compares well with the predictions.

5.7 Chapter Recap

As we conclude this rather long chapter, with many interesting concepts and hands-on techniques to grasp, let's spend a moment revisiting the key concepts and techniques that we have explored so far.

We began with morphological analysis and tokenization, where we probed into multiple tokenization techniques, including whitespace and MWE tokenizers. We also presented hands-on code demos for each. These code snippets and functions you can directly adapt for your NLP projects. This is true for all the code demos included in this book. We then delved into the fundamental building stones of text analysis, stemming, and lemmatization. Here also, our code demos involved important Python libraries for text processing like NLTK's WordNet, spaCy, and TextBlob. Finally, we stitched together everything in a comprehensive lexical analysis with data preprocessing and predictive analysis project using YouTube Comments spam detection.

Lexical analysis is faced with several challenges that include complexities of natural language, diverse language structures, and evolving language patterns. Handling ambiguous words with multiple context-dependent meanings is especially difficult to handle and interpret. This uncertainty frequently leads to errors in the tasks like POS tagging. However, modern language models are able to resolve it up to a large extent. Another challenge is posed by a variety of language structures and the evolving nature of natural languages, including slang, idioms, domain-specific jargon, and evolving language patterns in digital communication with the constant evolution of new words and phrases. New sophisticated models are required to better handle these complex linguistic patterns. Researchers are working on newer models for real-time lexical analyzers and the development of better tools for multilingual and cross-domain text processing.

As you read further through this book, keep in mind that the foundational concepts and tools learned in this chapter are pivotal for mastering advanced NLP applications. Whether you are building a simple spam filter, working on a sentiment analysis model, or working on any other advanced NLP project, the techniques and concepts that you have learned in this chapter can act as your pivotal toolkit.

5.8 References

[1] Alberto, T.C. and Lochter, J.V. (2017). YouTube Spam Collection. UCI Machine Learning Repository. https://doi.org/10.24432/C58885

[2] Cortinhas, S. (2022). *NLP3: Bag-of-Words and Similarity*. Available at: https://www.kaggle.com/code/samuelcortinhas/nlp3-bag-of-words-and-similarity, Last Accessed: August 17, 2024.

CHAPTER 6

Syntactic and Semantic Techniques in NLP

6.1 Why You Should Read This Chapter

This chapter introduces some foundational concepts in syntactic analysis, like breaking down and analyzing sentence structures. Understanding these concepts is critical for advancing your skills in NLP. Knowledge of part-of-speech (POS) tagging and parsing techniques will help you in building better NLP models. POS tagging is an essential skill for constructing chatbots, search engine optimization (SEO), and machine translation. This chapter is full of practical examples, including a full-length solved tutorial to provide you with hands-on skills in this art. Skills in syntactic analysis enhance NLP systems, enabling them to grasp the basic structure of language, resulting in more nuanced and contextually accurate responses in applications such as virtual assistants and customer service chatbots.

Additionally, the skills learned in this chapter will prepare you to adapt and innovate with the latest developments in the NLP field. Before we take a deep dive into the topics of syntactic and semantic analysis, it is beneficial to formally understand them in simple words and comprehend the relationship between them. Finally, this section explains why you should master these topics.

Syntactic and semantic analysis are vital components of NLP. The main thrust of **syntactic analysis** is on the grammatical structure of sentences to make sure they follow the prescribed rules of syntax. The syntactic analysis breaks down the input sentences into fragments, such as phrases and clauses, and signifies their hierarchical relationships by utilizing structures like parse trees. For example, the sentence "The dog barked loudly" classifies "The dog" as a noun phrase and "barked loudly" as a verb phrase. It helps NLP application to comprehend how words (in a given sentence) relate to each other structurally. Syntactic analysis is crucial for NLP tasks, such as grammar checking, sentence parsing, and transforming unstructured text into a format that machines can read and understand.

Semantic analysis, on the other hand, explores the proper meaning of words, phrases, and sentences of the input text. This technique infers relations between words and resolves uncertainties in the meaning of the same words in different contexts. For example, the word "bank" can have different meanings as a financial institution or as the edge of a river. Semantic analysis also explores the logical representations of text. Semantic analysis techniques, such as word sense disambiguation (WSD), named entity recognition, and semantic role labeling, enable NLP applications to extract the meanings of words and phrases within their specific contexts. For instance, semantic analysis techniques can explain whether "Flying planes can be dangerous" denotes the act of flying planes or the planes (as machines) themselves as risky. Semantic analysis is critical for NLP systems such as social media or email sentiment analysis, question-answering chatbots, and machine translation from one language to another.

CHAPTER 6 SYNTACTIC AND SEMANTIC TECHNIQUES IN NLP

Syntactic and semantic analysis techniques are complementary to each other. The first ensures grammatical accuracy and structural consistency. The second (semantic analysis), on the other hand, ensures meaningful interpretation of words, phrases, and sentences in any given input text in NLP applications. A syntactically binding sentence might still be ambiguous from the semantic angle without proper context. When used together, both of these techniques enhance the precision and correctness of NLP applications. The correct application of both these techniques ensures proper structural correctness with proper contextual meaning.

Learning both these topics is critical for anyone working with NLP systems. These techniques form the backbone of language applications, which require grammatical correctness and proper contextual understanding. A proper understanding of syntactic and semantic techniques can help you in designing robust language models that are capable of parsing text accurately, along with proper interpretation of contextual meaning. This knowledge is crucial for developing cutting-edge AI systems, such as advanced domain-specific chatbots, virtual assistants like Alexa, and automated text summarization applications that can interact seamlessly with human languages, including English, German, and Hindi.

6.2 An Overview of Structural Techniques

Key concepts covered in the chapter, such as POS tagging, parsing, and syntax trees, are fundamental to understanding the structure and meaning of sentences. POS tagging finds the role of each word in a sentence. It enables a more accurate interpretation of text that is critical in many NLP tasks, such as grammar correction.

Parsing techniques such as dependency and constituency parsing enable the breaking down of sentences into their respective grammatical components. It facilitates a more precise analysis of word relationships, which is crucial for accurately extracting information from input text.

Syntax trees are a visual way to represent sentence structures. They provide a clear overview of how words are organized in any input sentence. Syntax trees are helpful in text processing tasks, such as syntactic parsing, which enhances the effectiveness and reliability of NLP models in real-world scenarios, including question-answering systems, text summarization, and speech recognition.

Our discussion in this section would not be complete without taking a couple of business or industry applications in this context. Let's case the real-time application like machine translation: syntactic analysis is useful here to accurately parse the input sentences in the source language to identify the underlying grammatical structures and dependencies between words. This information helps in NLP models to generate grammatically correct translations in the target language while preserving the original sentence structure. Information extraction systems in many NLP applications use syntactic analysis to parse sentences and extract key entities, relationships, and actions. In another real-life scenario of legal contracts, syntactic parsing here can identify and structure important legal terms along with their connections. It's helpful in the automated extraction of relevant clauses and obligations. Next, let's examine the key concepts of this chapter one by one and try to develop the conceptual and hands-on skills at each one of them.

6.2.1 Part-of-Speech Tagging

Part-of-speech (POS) tagging is a fundamental step in greater NLP tasks. It labels each word in a sentence with its corresponding role within the context of the sentence. These roles can be nouns, verbs, adjectives, or adverbs. POS tagging helps us understand the grammatical structure of a sentence.

POS tagging lays the foundation for more advanced NLP tasks, such as parsing, information extraction, and text summarization. The POS tagging step typically uses algorithms that analyze the context of words by leveraging statistical models or rule-based systems to assign the correct tags.

POS tagging can be resourcefully done using popular NLP libraries like NLTK, spaCy, and StanfordNLP. These libraries use pre-built models and tools that simplify the process of POS tagging. NLTK is ideal for educational purposes and comparatively simpler projects. spaCy, on the other hand, is known for its speed and accuracy. It is popularly used in production environments. spaCy also provides advanced features such as dependency parsing along with POS tagging, which makes it suitable for complex tasks. StanfordNLP is another powerful library offering high-accuracy POS tagging through its deep learning–based models. It supports multiple languages, making it a versatile choice for diverse NLP applications. All these libraries load a pre-trained model and pass a sentence or document through the model to get the POS tags for every word in an input sentence. The ease of use in these libraries allows NLP developers to integrate POS tagging into larger NLP pipelines to increase the text processing capabilities of their applications.

In the English language, words fall into one of eight or nine parts of speech as follows.

- Noun
- Verb
- Adjective
- Adverb
- Pronoun
- Preposition
- Conjunction
- Interjection
- Determiner

CHAPTER 6   SYNTACTIC AND SEMANTIC TECHNIQUES IN NLP

We do not explain each POS tag. You are encouraged to use any English language grammar books. The nine POS tags are the most basic ones. In practice, the list of POS tags is much larger. For example, Table 6-1 features the POS tags used by a CMU tagger. This tagger uses a total of 37 POS tags, as listed in Table 6-1. A CMU tagger is a part-of-speech tagging tool specially designed for accurate and efficient POS tagging. Carnegie Mellon University developed it.

Table 6-1. POS Tags Used by CMU Tagger

	Tag	Description		Tag	Description
1.	' '	Quotation mark (")	20.	RB	Adverb
2.	,	Comma	21.	RBR	Adverb, comparative
3.	.	Period (. ? !)	22.	RBS	Adverb, superlative
4.	:	Punctuation (: ; … + − = < > / [ ] ~)	23.	RP	Particle
5.	CC	Coordinating conjunction	24.	RT	Retweet
6.	CD	Cardinal number	25.	TO	*to*
7.	DT	Determiner	26.	UH	Interjection
8.	EX	Existential *there*	27.	URL	Universal Resource Locator
9.	HT	Hashtag	28.	USR	Username (preceded by @)
10.	IN	Preposition or subordinating conjunction	29.	VB	Verb, base form
11.	JJ	Adjective	30.	VBD	Verb, past tense
12.	JJR	Adjective, comparative	31.	VBG	Verb, gerund or present participle
13.	JJS	Adjective, superlative	32.	VBN	Verb, past participle
14.	MD	Modal	33.	VBP	Verb, non-3rd person singular present
15.	NN	Noun, singular or mass	34.	VBZ	Verb, 3rd person singular present
16.	NNP	Proper noun, singular	35.	WDT	Wh-determiner
17.	NNS	Noun, plural	36.	WP	Wh-pronoun
18.	PRP	Personal pronoun	37.	WRB	Wh-adverb
19.	PRP$	Possessive pronoun			

Table 6-1 lists 37 POS tags; such a large number is required to capture the nuanced grammatical distinctions in language. More accurate parsing of sentences requires this level of detail. It allows the understanding of complex sentence structures by differentiating between similar but distinct word roles. Figure 6-1 depicts a simple example of POS tagging.

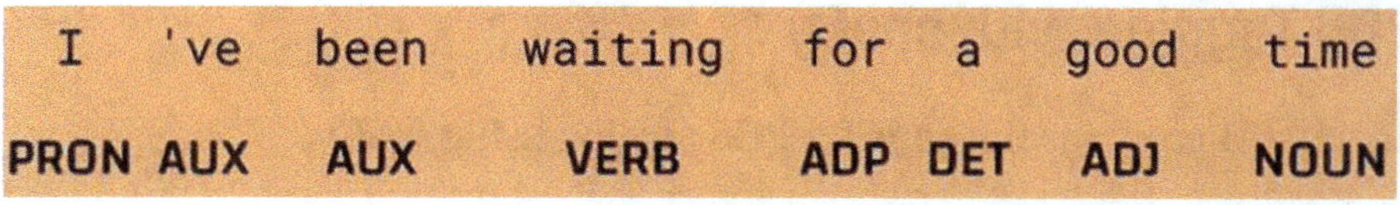

CHAPTER 6 SYNTACTIC AND SEMANTIC TECHNIQUES IN NLP

I	've	been	waiting	for	a	good	time
PRON	AUX	AUX	VERB	ADP	DET	ADJ	NOUN

Figure 6-1. *A simple example of POS tagging*

6.2.2 POS Tagging Tutorials

The following are coding tutorials for POS tagging with NLTK and spaCy.

POS Tagging with NLTK

```
!pip install nltk # Do it, if you gve not done lt already.
```

import nltk
nltk.download('punkt') # *This we have seen n prevous chapters.*
nltk.download('averaged_perceptron_tagger') # *Tagger based on averaged perceptron algorithm.*

```
sentence = "nltk it ideal for educational purposes and comparatively simpler projects."
tokens = nltk.word_tokenize(sentence)
pos_tags = nltk.pos_tag(tokens)
print(pos_tags)
```

OUTPUT:

[('nltk', 'IN'), ('it', 'PRP'), ('ideal', 'VB'), ('for', 'IN'), ('educational', 'JJ'), ('purposes', 'NNS'), ('and', 'CC'), ('comparatively', 'RB'), ('simpler', 'JJ'), ('projects', 'NNS'), ('.', '.')]

CHAPTER 6   SYNTACTIC AND SEMANTIC TECHNIQUES IN NLP

POS Tagging with spaCy

Listing 6-1 contains the code for POS tagging using spaCy.

Listing 6-1. POS tagging

```
!pip install spacy # Do it, if you gve not done lt already.
# "en_core_web_sm" is a small, lightweight English language model in spaCy.
# It is ideal for quick NLP tasks.
!python -m spacy download en_core_web_sm

import spacy
nlp = spacy.load("en_core_web_sm")

Sample_doc = nlp("""Spacy, on the other hand, is known for its speed and accuracy. It is popularly used in production environments. """)
for token in Sample_doc:
    print(token.text, token.pos_)
```

OUTPUT:

```
Spacy NOUN
, PUNCT
on ADP
the DET
other ADJ
hand NOUN
, PUNCT
is AUX
known VERB
for ADP
its PRON
speed NOUN
```

CHAPTER 6   SYNTACTIC AND SEMANTIC TECHNIQUES IN NLP

and CCONJ
accuracy NOUN
. PUNCT

   SPACE
It PRON
is AUX
popularly ADV
used VERB
in ADP
production NOUN
environments NOUN
. PUNCT

Note The StanfordNLP demo is not included as its dependencies installing was taking much longer than reasonably expected.

Many NLP professionals use more complex LSTM-based POS tagging or other deep learning methods for the same because they may offer better performance for complex or specialized tasks. spaCy provides a faster and more user-friendly POS tagging approach, so it is better suited for r general-purpose applications. The choice of POS tagging technique depends upon the specific needs of your project. Stay tuned!

We close this section on POS tagging by putting forward their immediate uses, which you will see in this book.

- POS tagging plays an important role in building lemmatizes, which are used to reduce a word to its root form.

- POS tags are useful for building parse trees and extracting relations between words, which are discussed in the next section.

- POS tags find use in NERs. Examples of NER are names of people, organizations, locations, and dates. Keep in mind that most named entities are Nouns.

6.2.3 Parsing Techniques

Parsing is breaking down a sentence into its constituent components and analyzing the relationships between words. Parsing allows machines to infer and process human languages more effectively. Parsing techniques provide the required framework for understanding the grammatical structure of sentences. Parsing techniques are divided into constituency parsing and dependency parsing. Constituency parsing finds the structure of a sentence by dividing it into nested phrases or constituents. Dependency parsing, on the other hand, keeps its focus on the relationships between words. It determines how each word depends on another within the sentence. With a proper understanding of how sentences are structured and how words relate to each other, NLP developers can produce systems that can better capture the delicacies of human language for building effective NLP solutions. Next, let's look at dependency and constituency parsing in the details, just enough to understand them. Then, we go through a hands-on example.

Dependency Parsing

Let's start with an example sentence, "She gave the book to him." A dependency parser breaks it down by detecting the grammatical relationships between the words and organizing them into a tree-like structure. This is how it works.

1. Root: The parser identifies the main verb "gave", which gives us the root of the sentence.

2. Subject (nsubj): The parser identifies "She" as the subject of the verb "gave."

3. Direct object (dobj): Next, the parser detects "book" as the direct object of "gave."

4. Prepositional phrase (prep): The parser finally characterizes "to him" as a prepositional phrase.

At the end, the dependency parser cuts the following structure to visually represent the grammatical structure of the given input sentence "She gave the book to him."

- **"gave"** (root)
  - **"She"** (nsubj)
  - **"book"** (dobj)
  - **"to"** (prep)
    - **"him"** (pobj)

Common tags used in dependency parsing are as follows:

- **nsubj** (nominal subject): The noun or noun phrase, which is the subject of the verb.

- **dobj** (direct object): The noun or noun phrase, which directly receives the action of the verb.

- **iobj** (indirect object): The noun or noun phrase, which indirectly receives the action of the verb.

- **attr** (attribute): An adjective or adjective phrase that describes a noun.

- **amod** (adjectival modifier): An adjective modifying a noun.

- **advmod** (adverbial modifier): An adverb modifying a verb, adjective, or another adverb.

- **prep** (prepositional modifier): A preposition introducing a prepositional phrase.

- **pobj** (prepositional object): The noun or noun phrase that follows a preposition.

- **det** (determiner): A word like "the" or "a" that modifies a noun.

- **conj** (conjunction): A word that connects other words or phrases, like "and" or "but".

- **cc** (coordinating conjunction): A word that connects words or phrases of the same type.

- **nmod** (nominal modifier): A noun that modifies another noun.

- **aux** (auxiliary verb): A verb that helps form different tenses, moods, or voices of the main verb.

- **expl** (expletive): A word like "there" that does not have a substantive meaning but serves a grammatical function.

This method (dependency parsing) helps machines (read computers) realize input sentences by breaking them down into the connections depicted in the preceding tree. This process makes it easier for computers to figure out the meaning of the sentence. This way, instead of just recognizing the words, machines learn how the words fit together and work together in a sentence.

Dependency parsing involves breaking down an input sentence into the word level and then identifying the grammatical relationships between those words. In short, dependency parsing shows how each word connects to others by detecting the sentence structure.

CHAPTER 6   SYNTACTIC AND SEMANTIC TECHNIQUES IN NLP

Constituency Parsing

The constituency parsing technique builds a hierarchical tree that depicts how words group together into phrases. Dependency parsing, on the other hand, involves breaking down sentences to the word level.

Let's take an example sentence, "The cat sat on the mat."
First, identify the phrases.

- "The cat" is a noun phrase (NP) that serves as the subject.
- "sat on the mat" is a verb phrase (VP) that tells about the action and its details.
- "on the mat" is a prepositional phrase (PP) that modifies the verb.

Next, build the tree.

- The main verb phrase "sat on the mat" is at the root of the tree.
- "The cat" will serve as the subject.
- "on the mat" is connected to the verb phrase as an additional detail.

The formal tree resembles the following.

```
(S
  (NP The cat)
  (VP
    (V sat)
    (PP
      (P on)
```

```
      (NP the mat)
    )
  )
)
```

In this tree,

- S represents the entire sentence.
- NP for noun phrases.
- VP represents verb phrases.
- PP stands for prepositional phrases.

The following are common tags used in constituency parsing.

- **S** (sentence): Stands for the complete sentence.
- **NP** (noun phrase): It groups nouns and their modifiers.
- **VP** (verb phrase): It stands for the main verb and its arguments.
- **PP** (prepositional phrase): Includes a preposition and its object.
- **AdjP** (adjective Phrase): It contains adjectives and their modifiers.
- **AdvP** (adverb phrase): It groups adverbs and their modifiers.
- **Det** (determiner): It stands for the words like "the," "a," or "an" that modify nouns.
- **N** (noun): Stands for nouns.
- **V** (verb): Represents verbs.
- **P** (preposition): Stands for prepositions.

Constituency parsing and dependency parsing use different tags because they serve different purposes. POS tags are also different from the tags used in constituency and dependency parsing

6.2.4 Hands-on Example Covering for Parsing

Constituency parsing and dependency parsing are two different approaches to parsing. The first one emphasizes hierarchical structure, while the second focuses on grammatical relations. Which one has more prevalence in practice? Dependency parsing is more prevalent in practice. The reason is as follows. Dependency parsing is often aligned with the needs of many NLP tasks as its focus is on grammatical relationships, which plays a pivotal role in understanding sentence meaning. Dependency parsing is also generally more computationally efficient and flexible across languages. For these reasons, the code demo in the following section only uses a dependency parser.

Dependency Parsing Code Demo

Listing 6-2 and Figure 6-2 are examples of dependency parsing.

Listing 6-2. Dependency Parcing

```
"""
You are strongly advised to run the following conda commands
on Anaconda command prompt before attempting to install or
import spaCy.
"""

conda install pytorch torchvision torchaudio pytorch-cuda=11.8 -c pytorch -c nvidia
conda install pytorch torchvision torchaudio cpuonly -c pytorch
```

CHAPTER 6   SYNTACTIC AND SEMANTIC TECHNIQUES IN NLP

```python
import spacy
# Displacy is SpaCy's visualization tool for rendering
syntactic dependency parses in HTML format.
from spacy import displacy

# Load the required pre-trained SpaCy model.
nlp = spacy.load('en_core_web_sm')

# Create a sample sentence for processing.
sample_sentence = "She sells seashells by the seashore."
doc = nlp(sample_sentence)

# Print the dependency parsing result
for token in doc:
    print(f"Token: {token.text}, Head: {token.head.text},
    Dependency: {token.dep_}")

# Set up options for visualization (fit in page).
options = {
    "compact": True,  # Makes the visualization more compact.
    "color": "blue",  # Changes color of the dependency lines.
    "bg": "white",    # Background color.
    "font": "Source Sans Pro",  # Font style.
    "arrow_stroke": 2,  # Thickness of the arrows.
    "arrow_stretch": 8,  # Stretch of the arrows.
    "distance": 100     # Distance between tokens.
}

# Render the dependency parser with the options set above.
displacy.render(doc, style='dep', options=options,
jupyter=True)
```

OUTPUT:

Token: She, Head: sells, Dependency: nsubj
Token: sells, Head: sells, Dependency: ROOT

CHAPTER 6 SYNTACTIC AND SEMANTIC TECHNIQUES IN NLP

```
Token: seashells, Head: sells, Dependency: dobj
Token: by, Head: sells, Dependency: prep
Token: the, Head: seashore, Dependency: det
Token: seashore, Head: by, Dependency: pobj
Token: ., Head: sells, Dependency: punct
<IPython.core.display.HTML object>
```

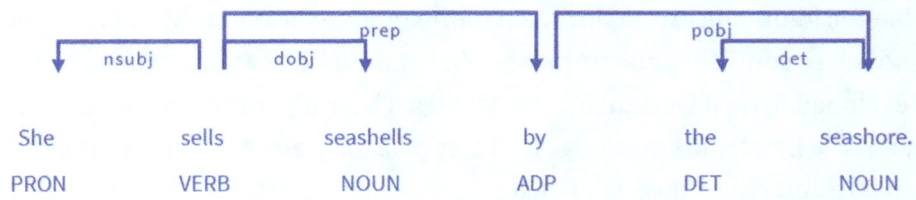

Figure 6-2. *Dependency tree for the sample, "She sells seashells by the seashore."*

The following are explanations.

- Root node ("sells"): The verb "sells" is the central action. It forms the root of the tree.

- Subject ("She"): "She" is the subject. It performs the action of selling.

- Direct object ("seashells"): "Seashells" is the object being sold. It is linked directly to the verb.

- Prepositional phrase ("by the seashore"): "By" introduces the location. It modifies the verb "sells."

- Modifiers ("the"): "The" specifies the noun "seashore." It acts as a determiner.

6.2.5 Challenges in Syntactic Analysis

Input sentences can often be understood in several ways. The interpretation depends upon the context, word order, or even cultural nuances. So, it is always challenging to construct parsers that are able to maintain consistency in terms of producing accurate syntactic structures for complex sentences. Complex sentence structures can include nested clauses, idiomatic expressions, and non-standard grammar. All of it can be enough to confuse syntactic parsers. Several sophisticated algorithms are developed to deal with such complexities. These algorithms have built-in extensive linguistic knowledge and complex mathematics to accurately analyze and crack these challenges.

The issue of resource limitations is an additional roadblock that syntactic analyzers have to deal with. High-quality syntactic parsing typically needs huge computational resources and large labeled datasets. Both, the computing resources and labeled datasets, are costly and time-consuming to acquire. When it comes to different languages or specialized domains, syntactic parsers often require significant customization. It further adds to the complexity in the way of effective syntactic analysis for diverse NLP applications. Let's try to understand this with a sample sentence as follows.

"He told his friend that he will win the race."

You might have guessed the ambiguity by now. The pronoun "he" could refer to either "he" or "his friend," leading to two possible interpretations:

- He (the original subject) will win the race.
- His friend will win the race.

The parser needs to take the help of context or some extra information to resolve the disambiguate to correctly infer who is expected to win the race. It could either be the speaker or the speaker's friend. It is dependent on the context.

CHAPTER 6 SYNTACTIC AND SEMANTIC TECHNIQUES IN NLP

We can take the help by generating several parse trees for the same sentence. Every tree can represent a different possible structure. Each structure depicts how the phrases and clauses relate to each other. The parser then uses grammatical rules to assess each of these trees and finally resolve the most coherent structure.

Parsers also employ statistical models and special disambiguation algorithms to further refine their analysis. Such statistical models are already trained on large relevant datasets. These models can predict (with a probability) which interpretation is most likely based on historical usage patterns. A probabilistic context-free grammar is also employed to compare several parse trees and select the most probabilistic interpretation.

More sophisticated parsers use a combination of these techniques, using context, grammatical rules, and statistical data to effectively handle ambiguities and resolve issues related to complex sentences.

6.3 Introduction to Semantics

Semantic analysis concentrates on understanding the meaning of words, sentences, and entire texts. In lexical analysis, the focus is on breaking down text into words or basic units of tokens. And syntactic analysis involves the structure and grammar of sentences. Unlike syntactic analysis, semantic analysis probes into the understanding of the meaning (of words and sentences) so that machines can comprehend human language in a contextually accurate and meaningful manner. Under semantic analysis, we examine a few key concepts, including chunking and term frequency–inverse document frequency, which are classically linked to syntactic techniques.

Word sense disambiguation (WSD) is a pivotal technique in semantic analysis. It gets us the correct meaning of a word based on its context. For instance, consider a sample sentence, "The Bank of London is situated on the east bank of the River Thames." Here, the word "bank" could refer to a financial institution or the side of a river. WSD helps in resolving such ambiguities and finds the appropriate context-based meaning of a word(s) in a given input sentence.

The semantic similarity technique is used to find out how similar or different two pieces of text are in terms of meaning. It helps in understanding the meaning and context of the text, allowing documents with similar themes or topics to be clustered accurately, a process known as document clustering.

NER techniques identify and classify entities in text. Entities are names of people, organizations, locations, dates, and so forth. So, in a way, NER extracts structured information from unstructured text.

Chunking or shallow parsing focuses on breaking sentences into smaller, syntactically related chunks such as noun or verb phrases. It helps in understanding the structure of a sentence by breaking it down into its component parts.

A term-document matrix (or co-occurrence matrix) captures the frequency of various terms across the input documents. It is concerned with how often words appear. Term frequency–inverse document frequency (TF–IDF) is built on a term-document matrix because it further hones the importance of these terms. TF–IDF considers both the frequency of terms (in a given input document) and their rarity across the entire input corpus. TF–IDF offers a more nuanced understanding of word significance.

These techniques help bridge the gap between words and their meaning to build a deeper understanding of language.

6.3.1 Chunking in NLP

Chunking breaks down an input sentence into smaller, meaningful units, which can be as simple as noun phrases (NP) or verb phrases (VP). Grouping words into chunks helps to grasp the syntactic structure of a sentence. Chunking makes it simpler to analyze the roles played

CHAPTER 6 SYNTACTIC AND SEMANTIC TECHNIQUES IN NLP

by individual words. In chunking, the focus is on extracting essential information quickly (partial parsing). Full parsing, on the other hand, seeks to analyze the complete grammatical structure of input sentences. Chunking finds uses in specific tasks, such as NER and question-answering systems, in which quickly extracting specific phrases is more important than analyzing the entire grammatical structure of a sentence.

Let's apply chunking to the first sentence of this paragraph. The following repeats it for convenience.

"""

Chunking, as we talked earlier, is breaking down an input sentence into smaller, meaningful units, which can be as simple as noun phrases (NP) or verb phrases (VP). Grouping of words into chunks, helps to grasp the syntactic structure of a sentence.

"""

We can extract the following six chunks out of this sentence:

1. Chunking, (NP)
2. as we talked earlier, (PP: prepositional phrase)
3. is breaking down (VP: verb phrase)
4. an input sentence (NP: noun phrase)
5. into smaller, meaningful units, (PP)
6. which can be as simple as noun phrases (NP) or verb phrases (VP). (Clause)

In this example, noun phrases (NP), verb phrases (VP), and prepositional phrases (PP) have been recognized as the significant chunks of the sentence.

CHAPTER 6   SYNTACTIC AND SEMANTIC TECHNIQUES IN NLP

Let's take one more example as follows.

""

Full parsing, on the other hand, seeks to analyze the complete grammatical structure of input sentences.

""

We can extract the following five chunks out of this second example.

1. Full parsing, (NP: noun phrase)

2. on the other hand, (PP: prepositional phrase)

3. seeks to analyze (VP: verb phrase)

4. the complete grammatical structure (NP: noun phrase)

5. of input sentences. (PP: prepositional phrase)

Noun phrases (NP), verb phrases (VP), and prepositional phrases (PP) have been found as the key components (chunks) of the sentence.

With these examples, chunks and chunking might be pretty clear to you. If not, you can refer to any standard NLP book for additional examples.

For chunking, the algorithm(s) groups together syntactically related units (words) like noun phrases (NP) or verb phrases (VP). During chunking, the input text is skimmed through for these patterns. On finding a match, the corresponding words are grouped together to form a chunk. Chunking avoids a full syntactic analysis, but it still breaks the sentence into its most meaningful components. This makes chunking a fast and efficient technique to parse input text for further processing.

The foremost challenge in chunking is the ambiguity in language. Words have multiple roles depending on context. This makes it difficult to accurately recognize and group them into the correct chunks. Another challenge in chunking is dealing with exceptions and variations in sentence structure. Additionally, chunking finds it difficult to deal with out-of-vocabulary words and uncommon phrases. These challenges can be handled to some extent using models that employ a combination of rule-based and machine learning approaches.

NLTK, spaCy, and TextBlob are the popular Python libraries that professionals and researchers use for chunking. These standard libraries are convenient and powerful, but custom deep learning solutions may be required for scalability and specialized NLP tasks. Advanced models using transformer models can further enhance chunking capabilities. Such custom models may sometimes offer better precision, flexibility, and scalability. We include a code demo of chunking toward the end of this chapter. Stay tuned!

6.3.2 Named Entity Recognition

Named entity recognition (NER) focuses on recognizing and classifying named entities in text. Named entities are names of people, organizations, locations, dates, and more. Identifying named entities is crucial for understanding the context and meaning of the text.

Let's try to understand it by a couple of examples as follows.

In the sentence, "John shifted to New York on 24th August 2020 with his family," the NER would extract the named entities as John (person), New York (location), and 24th August 2020 (date).

In the sentence, "Sarah traveled to San Francisco on 10th October 2023 to attend a conference," the named entities are Sarah (person), San Francisco (location), and 10th October 2023 (date).

Note that "San Francisco" and "New Delhi" are multi-word or multi-token entities. NER programs handle such words by examining the words around them and recognizing patterns—specifically, contextual analysis and sequence labeling. Some NER systems have pre-compiled lists of known entities (gazetteers) to help detect such multi-word entities.

For detecting named entities, the entire text is scanned to identify the phrases that fit certain predefined categories like "person," "location," and "dates." The challenges in NER are similar to those in WSD, including language ambiguity, context understanding, variability in naming, incomplete data, and differences in language and dialect.

In applications such as news aggregation and summarization, NER facilitates the summarization of news articles and helps prepare news clusters around specific topics or entities. There are many other applications of NER, like customer service automation, financial market analysis, healthcare applications, and social media monitoring. NER helps service automatize customer service applications by identifying key entities in queries for precise responses. It supports financial market analysis, healthcare applications, and social media monitoring by mining relevant information for informed decision-making and efficient data management.

6.3.3 WSD Revisited

Let's revisit WSD's relatively unfamiliar aspects, starting with how it integrates with larger applications. WSD can further enhance the accuracy of NER and information extraction. Suppose the input text contains a sentence like "Apple released its new MacBook model at a press conference in Cupertino." As such, "apple" is recognized as a named entity (fruit), but WSD can correctly disambiguate it as a company by correctly analyzing its context. This type of accurate WSD can also improve the overall performance of other NLP systems. Correct word disambiguation can be useful in several other NLP tasks like machine translation, sentiment analysis, search engines, voice assistants, and text summarization. Accurate interpretation of ambiguous words enhances the effectiveness and results of these applications.

CHAPTER 6   SYNTACTIC AND SEMANTIC TECHNIQUES IN NLP

Now let's talk about a couple of more popular methods for performing WSD: knowledge-based, statistical, and neural methods. Knowledge-based methods employ predefined dictionaries and rules for WSD. They offer high accuracy but are limited in adaptability. Statistical models use probabilistic models and training on large corpora to forecast word meanings depending on context. Neural network models are powered deep learning to capture complex patterns and contextual nuances. They are useful for more dynamic and accurate disambiguation in complex scenarios, but they are more resource-consuming in terms of time, computing power, and extensive training data.

Knowledge-based methods are inflexible and have limited adaptability as they are limited by their dependency on pre-existing rules and dictionaries. Statistical methods, on the other hand, are more adaptable, but they require large datasets to be effective, which may not always be available. Statistical methods have less precision in handling nuanced contexts because of their dependency on probabilistic models. Deep learning–based methods are powerful, but like most other applications using this technology, they need significant computational resources and extensive training data. This makes them costly and complex to implement. So, they are used in only special and complex business cases. Some models for WSD use a combination of these methods. This is done to leverage their individual strengths to address their limitations.

Like most NLP tasks, WSD also comes with its own limitations. Polysemy, that is, words with multiple meanings, pose a significant challenge, especially when the correct meaning depends on delicate contextual evidence. Insufficiently annotated data (data sparsity) for training WSD models is another limitation. It can limit the NLP model's generalization and performance. Adaptation to multiple domains for a model trained on a specific domain, as well as the computational complexity of models, especially those belonging to deep learning technologies, adds to the limitations of WSD models.

CHAPTER 6 SYNTACTIC AND SEMANTIC TECHNIQUES IN NLP

Advanced WSD techniques are based on transformer models that are capable of capturing deeper semantic relationships.

6.3.4 Term-Document Matrix (Co-Occurrence Matrix)

A term-document matrix helps to convert text data into a structured, numerical format, which is essential for NLP models to process it further. The term-document matrix, also known as a co-occurrence matrix, is a matrix that represents the frequency of words occurring in a document set. In this matrix, rows typically represent unique words (terms), and columns represent different documents. Each cell in the matrix contains the frequency of occurrence of a specific word in a particular document.

Let's try to understand the term-document matrix with the following set of three simplified documents.

- Document 1: "Apple and banana are fruits."
- Document 2: "I like to eat apple pie."
- Document 3: "The banana pie is delicious."

The next task is to find the 13 unique words in these documents: apple, banana, and, are, fruits, I, like, to, eat, pie, is, delicious, the.

Now, we need to build a matrix with rows representing these unique words and columns representing the three documents. Each cell in the following matrix specifies how many times a term appears in that particular document. Note that the term "apple" appears in document 1 as well as in document 2. "Banana" appears in documents 1 and 3. No other term repeats. The cells of the matrix (see Table 6-2) get specified accordingly.

CHAPTER 6   SYNTACTIC AND SEMANTIC TECHNIQUES IN NLP

Table 6-2. *Term-Document Matrix*

Term	Document 1	Document 2	Document 3
apple	1	1	0
banana	1	0	1
and	1	0	0
are	1	0	0
fruits	1	0	0
I	0	1	0
like	0	1	0
to	0	1	0
eat	0	1	0
pie	0	1	1
is	0	0	1
delicious	0	0	1
the	0	0	1

The matrix shown in Table 6-2 helps you quantitatively analyze the occurrence and distribution of words across documents (in a set). It's used for several NLP tasks, such as identifying key terms, clustering similar documents, and forming the basis for further analysis.

Here is how we can use the matrix specific to Table 6-2. This matrix enables us to visualize how the three given documents are represented as numeric vectors.

- Document 1 vector: [1, 1, 1, 1, 1, 0, 0, 0, 0, 0, 0, 0, 0]
- Document 2 vector: [1, 0, 0, 0, 0, 1, 1, 1, 1, 1, 0, 0, 0]
- Document 3 vector: [0, 1, 0, 0, 0, 0, 0, 0, 0, 1, 1, 1, 1]

These documents are now represented as numeric vectors, which are then fed into NLP machine learning models to classify them into predefined categories, such as spam detection or topic classification. There are numerous other applications of these numeric vector representations of the original text documents. We used similar vectors in our Chapter 1 code demos.

6.3.5 Term Frequency–Inverse Document Frequency

There are two components in this term. Term frequency (TF) is concerned with how frequently a term appears in a specified document. Inverse document frequency (IDF) decreases the weight of terms that appear in multiple documents across the given input corpus. IDF gives importance to unique terms. More formally, TF-IDF is a statistical measure used to assess the importance of a term in a document relative to a collection of documents (input corpus). It finds widespread use in text-mining tasks to rank the relevance of documents based on a query. It assigns more weights to the words that are common in a particular document but rare across the input corpus. TF-IDF, thus, helps to identify the most significant words that best represent the content of a document.

Let's try to understand this phenomenon using the same set of documents as we used in the previous section. The following repeats it for a ready reference.

- Document 1: "Apple and banana are fruits."
- Document 2: "I like to eat apple pie."
- Document 3: "The banana pie is delicious."

The next step is to calculate term frequency (TF) as follows.

To calculate TF, find out the number of times a term appears in a document and divide it by the total number of terms in that document.

- TF for document 1
  - apple: 1/5
  - banana: 1/5
  - and: 1/5
  - are: 1/5
  - fruits: 1/5

- TF for document 2
  - apple: 1/6
  - like: 1/6
  - to: 1/6
  - eat: 1/6
  - pie: 1/6
  - I: 1/6

- TF for document 2
  - banana: 1/6
  - pie: 1/6
  - is: 1/6
  - delicious: 1/6
  - the: 1/6

Then, calculate inverse document frequency (IDF). It is calculated as the logarithm of the total number of documents divided by the number of documents containing the term, which is as follows.

CHAPTER 6 SYNTACTIC AND SEMANTIC TECHNIQUES IN NLP

- apple: log(3/2) ≈ 0.18
- banana: log(3/2) ≈ 0.18
- pie: log(3/2) ≈ 0.18
- and: log(3/1) ≈ 0.48
- are: log(3/1) ≈ 0.48
- fruits: log(3/1) ≈ 0.48
- like, to, eat, I: log(3/1) ≈ 0.48
- is, delicious, the: log(3/1) ≈ 0.48

Finally, calculate the TF-IDF for each document by multiplying the TF by the IDF for each term in each document.

- Document 1
  - apple: 1/5 * 0.18 = 0.036
  - banana: 1/5 * 0.18 = 0.036
  - and: 1/5 * 0.48 = 0.096
  - are: 1/5 * 0.48 = 0.096
  - fruits: 1/5 * 0.48 = 0.096
- Document 2
  - apple: 1/6 * 0.18 = 0.03
  - pie: 1/6 * 0.18 = 0.03
  - like, to, eat, I: 1/6 * 0.48 ≈ 0.08 each
- Document 3
  - banana: 1/6 * 0.18 = 0.03
  - pie: 1/6 * 0.18 = 0.03
  - is, delicious, the: 1/6 * 0.48 ≈ 0.08 each

CHAPTER 6 SYNTACTIC AND SEMANTIC TECHNIQUES IN NLP

Note the following.

- In document 1, "and," "are," and "fruits" have slightly higher TF–IDF scores as compared to "apple," "banana," and "fruits." It is because "and," "are," and "fruits" are less common across documents.

- In the remaining documents, the terms "apple" and "banana" have lower TF–IDF scores as they appear in multiple documents. Unique words such as "I," "like," "delicious," and "the" get slightly higher scores.

We can conveniently represent TF–IDF scores for each term in the documents as a table. Table 6-3 shows how the TF–IDF matrix would look for the three example documents.

Table 6-3. *TF–IDF Matrix*

Term	Document 1	Document 2	Document 3
apple	0.036	0.03	0.03
banana	0.036	0.03	0.03
and	0.096	0.08	0.08
are	0.096	0.08	0.08
fruits	0.096	0.08	0.08
like	0	0.08	0
to	0	0.08	0
eat	0	0.08	0
pie	0	0.03	0.03
I	0	0.08	0
is	0	0	0.08
delicious	0	0	0.08
the	0	0	0.08

Let's take a few notes from Table 6-3.

- Each cell represents the TF–IDF score of each term in the given set of three input documents.

- For example, in document 1, "apple" has a TF–IDF score of 0.036, which indicates its importance relative to the document and the corpus.

- Terms like "and," "are," and "fruits" have higher TF–IDF scores in document 1 because they are more significant to this document compared to the others.

- TF–IDF improves the precision and effectiveness of NLP applications as it focuses on unique and significant and significant terms as completed a term-document matrix.

6.3.6 Coding Tutorials for This Section

The following is the Python code tutorial on term-document matrixes, TF–IDF, chunking, NER, and WSD.

```
# Install these libraries, if not done already.
# !pip install nltk spacy sklearn
!python -m spacy download en_core_web_sm
import nltk
import spacy
import numpy as np
import pandas as pd
from sklearn.feature_extraction.text import TfidfVectorizer
# Load SpaCy model.
nlp = spacy.load("en_core_web_sm")
```

```python
# Sample Documents.
documents = [
    "Apple and banana are fruits.",
    "I like to eat apple pie.",
    "The banana pie is delicious."]
```

Chunking

```python
def chunking_example(doc):
    chunks = []
    for sent in doc.sents:
        for chunk in sent.noun_chunks:
            chunks.append(chunk.text)
    return chunks
```

Named Entity Recognition

```python
def ner_example(doc):
    entities = [(ent.text, ent.label_) for ent in doc.ents]
    return entities
```

Word Sense Disambiguation

```python
def wsd_example(word, context):
    # Simplistic approach for demonstration
    if word == "apple":
        if "pie" in context:
            return "The tech company"
        else:
            return "The fruit"
    return None
```

Term-Document Matrix

```python
# Create the Term-Document Matrix using raw counts.
from sklearn.feature_extraction.text import CountVectorizer
vectorizer = CountVectorizer()
X = vectorizer.fit_transform(documents)

# Create DataFrame for better visualization
df_term_doc = pd.DataFrame(X.toarray(), columns=vectorizer.get_feature_names_out())
print("Term-Document Matrix:\n", df_term_doc)
```

Term-Document Matrix:

	and	apple	are	banana	delicious	eat	fruits	is	like	pie	the	to
0	1	1	1	1	0	0	1	0	0	0	0	0
1	0	1	0	0	0	1	0	0	1	1	0	1
2	0	0	0	1	1	0	0	1	0	1	1	0

TF–IDF

```python
from sklearn.feature_extraction.text import TfidfVectorizer
import pandas as pd

# Example Documents
documents = [
    "Apple and banana are fruits.",
    "I like to eat apple pie.",
    "The banana pie is delicious."
]

# Create the TF-IDF Matrix
vectorizer = TfidfVectorizer()
X = vectorizer.fit_transform(documents)
```

# Create DataFrame for better visualization
df_tfidf = pd.DataFrame(X.toarray(), columns=vectorizer.get_feature_names_out())

print("TF-IDF Matrix:\n", df_tfidf)
TF-IDF Matrix:
```
        and     apple       are    banana  delicious      eat    fruits  \
0  0.490479  0.373022  0.490479  0.373022   0.000000  0.00000
0  0.490479
1  0.000000  0.373022  0.000000  0.000000   0.000000  0.49047
9  0.000000
2  0.000000  0.000000  0.000000  0.373022   0.490479  0.00000
0  0.000000

         is      like       pie       the        to
0  0.000000  0.000000  0.000000  0.000000  0.000000
1  0.000000  0.490479  0.373022  0.000000  0.490479
2  0.490479  0.000000  0.373022  0.490479  0.000000
```

- The discrepancy between the manually calculated TF–IDF values and those calculated by the program might arise from several factors. It may be due to the way TF–IDF values are computed or normalized. It is also possible that certain preprocessing steps or default settings in the libraries might lead to differences.

- But both manual and program calculated values should show the same terms as significant.

    - Doc 1 significant terms: "Apple," "banana"

    - Doc 2 significant terms: "apple," "pie"

    - Doc 3 significant terms: "banana," "pie"

```
# Process sample documents.
docs_spacy = [nlp(doc) for doc in documents]
print(docs_spacy)
```

[Apple and banana are fruits., I like to eat apple pie., The banana pie is delicious.]

```
# Try chunking.
# Uses SpaCy to extract noun chunks from each document.
chunks = [chunking_example(doc) for doc in docs_spacy]
print("Chunks:", chunks)
```

Chunks: [['Apple', 'banana', 'fruits'], ['I', 'apple pie'], ['The banana pie']]

```
# Try NER.
# Identifies named entities such as persons, locations, and organizations in the documents.
ner_results = [ner_example(doc) for doc in docs_spacy]
print("Named Entities:",ner_results)
```

Named Entities: [[('Apple', 'ORG')], [], []]

- No named entities were detected in documents 2 and 3. It is likely because "apple pie" and "banana pie" are not recognized as a named entity.

- It could be because of model limitations of the pre-trained spaCy model (en_core_web_sm). Larger models (en_core_web_md or en_core_web_lg), have more extensive training data. So, you can try with them. By using a larger model or adjusting the text, you may get better results.

Listing 6-3. Context-Based Interpretation

```
# Try WSD.
'''
A simplistic approach is used here to demonstrate
how "apple" might be interpreted based on context.
'''

wsd_results = [wsd_example("apple", doc.text) for doc in
docs_spacy]
print("WSD Results:", wsd_results)
WSD Results: ['The fruit', 'The tech company', 'The tech
company']
```

6.4 Lexical Semantics

In NLP, understanding the words and their relationships is covered in lexical semantics. This concept is related to determining the meaning of each individual word in the sentence. Lexical semantics also focus on the relation of words with each other using synonyms, antonyms, and alike. It is also concerned with the change in the meaning of words in different contexts. Lexical semantics helps in NLP tasks like WSD and machine translation. Lexical semantics plays a pivotal role in the development of algorithms that are suitable for processing the human language more correctly, which is very similar to human reasoning. This section discusses the key concepts of lexical semantics, such as synonyms, antonyms, homophones, and polysemy. The following descriptions are meaningful but concise.

6.4.1 Synonyms

Other than similar meanings, the synonyms have much more role to play in NLP. NLP supports the creation of flexible and accurate language models. For every NLP model, it is necessary to understand that individual words or phrases convey different ideas in different contexts.

The following example shows the usage of the word "bright," which can have different meanings in different situations.

- Context 1: " The sun is very "bright" today."
- Context 2: "Sarah is a "bright" student."

In context 1, "bright" means shiny, while in context 2, the word "bright" means intelligent.

Many practical NLP tasks, like automation applications dealing with customer interactions, require the correct recognition of synonyms in different contexts, as this could result in incomplete or inaccurate interpretations of sentences. Synonyms are also useful in expanding the training data. The available training data can be diversified by substituting the words with their synonyms to generalize the machine learning models. Whenever we come across imbalanced datasets, this type of data augmentation technique is very useful. Failing to correctly recognize synonyms in different contexts could result in incomplete or inaccurate interpretations of sentences for NLP tasks like automation applications dealing with customer interactions.

Voice assistants and chatbots are practical applications in which correct identification of synonyms is very critical for the proper response in situations where people use different words to mean the same thing, even if they use slightly different words in the new language. For example, if an English word like "tired" is translated into Spanish, the translator might choose between "cansado" or "agotado" depending on the situation. Both words mean tired, but they fit better in different contexts.

CHAPTER 6   SYNTACTIC AND SEMANTIC TECHNIQUES IN NLP

Another important aspect of synonyms is their ability to generate diverse text by using different words that convey the same meaning. This flexibility allows applications, especially those in creative writing or automated customer responses, to create more diverse and engaging content. Integrating a range of synonyms, NLP systems can produce text that feels more natural and less repetitive, thereby ensuring that the generated text aligns with the intended tone and style.

Another technique for handling context-based synonyms is WSD. In this technique, the analysis of context is performed to determine the correct sense of words so that the most appropriate synonym is selected. For this, computational methods are employed to quantify the semantic similarity measurement of two words in their meanings. Several NLP tasks, such as document clustering and retrieval, can be performed by these methods. Vector representation of words comes under advanced NLP models. These methods are more sophisticated and help in understanding the relationship between words (and synonyms) and making possible more sophisticated understanding and generation of language. Vector representation words are covered in an upcoming chapter. We effectively make use of such vectors to convert text (words) to numeric representations for use in our NLP models.

6.4.2 Antonyms

Antonyms and synonyms work together in NLP applications. Antonyms represent differences, while synonyms emphasize similarities. Antonyms are very crucial for any language as they support differentiating the delicate variances in meaning and context.

Identifying the opposing words supports the interpretation of the intensity, tone, and underlying implications of a text. A proper understanding of antonyms supports the full scope of sentiment, irony, and contrast, which are crucial to have a deeper meaning of the text. For example, if we realize that "happy" is the opposite of "sad," it helps NLP algorithms to capture the sentiment polarity in a given context. The main contribution of antonyms is in the in-text classification, WSD, and machine translation by appropriate selection of the correct meaning of a word based on its context.

As with synonyms, understanding context is essential for finding antonyms. Let's take a sample statement as follows.

"

> The light box was easy to carry, but when the room was dark,
> I turned on the light to see better.

"

"Light" appears in two places in this sample statement. In the first appearance, it refers to weight, meaning the box is not heavy (the formal anonym in this case would be "heavy"). In the second appearance, "light" refers to illumination, meaning a source of light used to brighten the room (antonym in this appearance," dark").

Without a proper understanding of language, in some cases, it is very difficult to detect antonyms. For machines, this kind of expertise is very difficult to achieve. Some simple use cases for the easy detection of antonyms are "full" and "empty". But due to the conceptual differences in the deeper meaning of words such as "freedom" and "constraint", the identification of antonyms requires a deeper understanding of context and connotation.

Linguistic changes and cultural variations affect the antonym connections. Over a period of time, words may become less common and may also acquire new meanings. The dynamic nature of language makes the task of accurately detecting the specific antonym more difficult when used in a variety of datasets and various circumstances it is used.

For the detection of antonyms, both statistical and dictionary-based approaches are used. Statistical methods are used on a large quantity of words to analyze the word usage and predict the probable antonyms based on context. Another approach to deal with this is a dictionary-based

CHAPTER 6   SYNTACTIC AND SEMANTIC TECHNIQUES IN NLP

method, which mainly depends upon the predefined lexical resources. These lexical resources also list antonyms and thus directly provide the ready reference of a pair of words.

6.4.3 Homophones

Homophones are the words with the same sound but have different meanings. They have different spellings. Due to their same pronunciations, the right meaning of homophones depends mainly on the context. This can sometimes be very challenging in NLP.

Homophones are words that sound the same but have different meanings. They frequently have dissimilar spellings. Homophones pose some unique challenges in NLP. Grasping the right meaning, in this case, is dependent solely on the context. For example, "pair" and "pear" both these words have the same pronunciations, but they refer to completely different things. "Pair" stands for a set of two objects, while "pear" means a type of fruit. The following are more common instances of homonyms.

- "Their," "There," and "They're"
- "To," "Too," and "Two"
- "Right" and "Write"

Understanding homonyms is especially important in tasks like speech recognition and text-to-speech conversion. In the preceding applications, the accurate identification of homophones is very critical for preserving the meaning. An example of this is "pair" and "pear". The two phrases "I need a pair of shoes" and "I need a pear of shoes" have an entirely different meaning. In the later sentence, "pear" looks like a wrong usage in the given phrase.

Regular NLP models must be able to handle homophones effectively. They do so by relying on contextual clues and surrounding words to crack the intended meaning. Context-based WSD techniques leverage large text corpora for their model training. It helps models differentiate homophones in various contexts and correctly interpret them. The elusive nature of homophones often necessitates sophisticated NLP models that can efficiently use contextual clues to differentiate between homophones.

6.4.4 Homographs

Homographs are words that have the same spelling but are different in meaning. In special cases, they even different pronunciations. For instance, the word "lead" refers to a specific metal, pronounced ("led"). "Lead" also refers to an act of guiding someone, pronounced "leed"). This unique property of homographs can cause uncertainty in both written and spoken language. The meaning homographs is resolved based on its context within a sentence or overall text. The following are a couple of common examples of homographs. Notice that same spelled words in all the three examples have different meanings depending upon the context and surrounding words within a sentence.

- "Bow": A gesture of bending forward or a weapon or ribbon.

- "Tear": A drop from the eye or the act of ripping.

- "Wind": Moving air or the act of twisting.

Dealing with homographs, like homophones, also requires a cautious analysis of surrounding words and general context to precisely detect the intended meaning. As the same spelling can lead to diverse interpretations, it makes it necessary for NLP applications to use advanced language models. Such advanced techniques typically involve correctly processing context-based word disambiguation.

Homographs are predominantly challenging in language tasks involving speech, like text-to-speech and speech recognition. In such applications, the correct pronunciation is the key to conveying the precise meaning.

6.4.5 Polysemy

In polysemy, a single word or a phrase has diverse meanings, which relate to each other by a common origin or context. In polysemy, the spelling of the word remains the same for all its meanings. Its diverse meanings can only be differentiated by context rather than by variations in spelling. Polysemy involves a single word with multiple related meanings. Homographs, on the other hand, deal with words that share the same spelling but have different, sometimes unrelated, meanings and pronunciations. In other words, polysemous words have multiple meanings originating from the same origin, while homograph words have different origins and multiple meanings.

The following are a few examples.

- Let's consider the word "head." It can mean the upper part of the body or the leader of a group. It is a polysemous word because its meanings are related to the common concept of being at the forefront or in a leading position.

- Now consider the word "lead". It can represent a type of metal or guide. "Lead" is a homograph, not polysemy, because its meanings are unrelated and have different pronunciations.

An accurate understanding of polysemous words is crucial for many applications like translating text and sentiment analysis. As applicable in many other phenomena explained in this section, for correct interpretation of polysemous words, understanding of context is an absolute must for the language models.

6.4.6 Hyponyms

We open this section with an example. The word "vehicle" is a hypernym with hyponyms like "car," "truck," "bus," and "motorcycle." Hyponyms can sometimes have their own hyponyms. It creates an intricate network of relationships. More formally, hyponymy is a type of semantic relationship between words where one word (the hyponym) is a more specific instance of another word (the hypernym). In a simpler example, "cat" is a hyponym of the word "animal," as it is a specific type of animal. Hyponyms can specifically help language generation models in producing more diverse and detailed content, avoiding repetition and adding depth to descriptions. For example, in language generation tasks, instead of repeatedly using the general term "vehicle," a generative application can use hyponyms like "car," "truck," "bicycle," or "motorcycle" to add variety and specificity to the content. This can create more engaging text while providing more accurate information to readers. This way, the generated language can capture the richness of the original text. Other applications of homonyms can be dialogue-based applications, in which maintaining a natural and diverse conversation is critical.

Detecting homonyms in a given input text again involves analyzing the context in which a word is placed. Methods to detect homonyms can yet again be both statistical and dictionary-based. Statistical methods often analyze the frequency and co-occurrence of words in large text corpora to deduce various context-based meanings. In contrast, dictionary-based techniques employ lexical databases like WordNet to map words to their several senses. Both the techniques can as well complement each other. Statistical techniques are better at providing context-based clues, while dictionary-based methods offer precise definitions and sense distinctions.

6.4.7 Building Lexical Semantic Models with Code Examples

Now, let's put these concepts into action. We address a few NLP-related business problems linked to content personalization, information retrieval, and user engagement. Let's examine how each concept in this section contributes to solving real-world business problems.

- **Business problem 1**: Improve search relevance and content matching.

    - **Business solution**: Use synonyms to solve this problem. We try to detect the different ways (different words or phrases) that users can use for search. It ensures that relevant content is retrieved even if the exact query phrases are different from the wordings of the content.

    - **Technical solution**: Use `wn.synsets(token)` to retrieve synsets (synonyms) of a word and collect synonyms from the lemmas of these synsets.

- **Business problem 2**: Clarify content context and preventing misinterpretation.

    - **Business solution**: Use antonyms here to ensure that content is accurately understood in the given context.

    - **Technical solution**: Use `wn.synsets(token)` to detect antonyms by examining if the lemma has any antonyms.

- **Business problem 3**: Avoid errors in speech recognition and text analysis.

- **Business solution**: Use homophones in this case. It helps in confirming that the generated text or interpreted text does not hold errors due to phonetic similarities.

- **Technical solution**: WordNet does not provide any direct phonetic information. However, we can use lemma names to approximate homophones.

- **Business problem 4**: Ensure accurate text analysis and context understanding.

    - **Business solution**: Employ detecting homographs in this case as it helps in correctly interpreting text.

    - **Technical solution**: Use `wn.synsets(token)` to check different POS for the same word to recognize homographs.

- **Business problem 5**: Enhancing content understanding and contextual relevance.

    - **Business solution**: Detecting polysemy helps to improve the accuracy of content analysis tools.

    - **Technical solution**: Count the number of `synsets` for a word to identify if it has multiple meanings.

- **Business problem 6**: Categorizing and personalizing content.

    - **Business solution**: Recognizing hyponyms permits businesses to classify content more effectually and personalize user experiences by matching individual user interests with relevant content.

    - **Technical solution**: Employ hyponyms of generic words to match and classify specific tokens in the text.

CHAPTER 6   SYNTACTIC AND SEMANTIC TECHNIQUES IN NLP

Next, let's put these solutions into action using nltk functions.

Tutorial on Synonyms, Antonyms, Homophones, Homographs, Polysemy, and Hyponyms

```
# Imports the nltk library for text processing
import nltk

# Imports WordNet to access synonyms, antonyms, and word
meanings.
from nltk.corpus import wordnet as wn

# Imports the function to tokenize text into words.
from nltk.tokenize import word_tokenize
# Downloads the WordNet lexical database.
nltk.download('wordnet')

# Downloads the Open multilingual WordNet package.
# It is needed for some language-related tasks.
nltk.download('omw-1.4')

# Downloads the Punkt tokenizer models for sentence and word
tokenization.
nltk.download('punkt')
# Mention the sample text.
article_text = """
The lead engineer in the new bridge project has made a
breakthrough.
The team is now looking at the potential impacts of their
findings on the new soil.
Meanwhile, a local cricket player hit his double ton in sports
news last night.
The lead engineer of the project team is highly respected.
"""
```

Tokenize the Article

```
# Tokenizes the article text into individual words or tokens.
tokens = word_tokenize(article_text)
```

Synonyms Detection

```
# Prints the header for synonym detection.
print("Synonyms Detection:")

# Iterates over the first 5 tokens.
for token in tokens[:5]:
    # Retrieves WordNet synsets for the token.
    synsets = wn.synsets(token)

    # Initializes a set to collect synonyms.
    synonyms = set()

    # Iterates over each synset to gather synonyms.
    for synset in synsets:
        for lemma in synset.lemmas():
            synonyms.add(lemma.name())

    # Prints the synonyms if found, otherwise, it indicates
      none were found.
    if synonyms:
        print(f"Synonyms for '{token}': {', '.join(synonyms)}")
    else:
        print(f"No synonyms found for '{token}'")
```

Synonyms Detection:
No synonyms found for 'The'
Synonyms for 'lead': wind, steer, lead-in, Pb, lede, jumper_lead, extend, trail, guide, moderate, head, booster_cable, precede, tip, result, jumper_cable, confidential_information,

pass, principal, atomic_number_82, run, track, lead_story, take, pencil_lead, leading, tether, contribute, star, hint, direct, conduct, leave, conduce, lead, top, go, spark_advance, leash, chair Synonyms for 'engineer': organize, mastermind, engineer, technologist, direct, orchestrate, applied_scientist, railroad_engineer, locomotive_engineer, engine_driver, organize Synonyms for 'in': inch, IN, Indiana, In, inward, inwards, in, indium, atomic_number_49, Hoosier_State
No synonyms found for 'the'

Antonyms Detection

```
# Prints the header for antonym detection.
print("\nAntonyms Detection:")

# Iterates over each token.
for token in tokens:
    # Retrieves WordNet synsets for the token.
    synsets = wn.synsets(token)

    # Initializes a set to collect antonyms.
    antonyms = set()

    # Iterates over each synset to gather antonyms.
    for synset in synsets:
        for lemma in synset.lemmas():
            # Checks for antonyms and adds them to the set.
            if lemma.antonyms():
                antonyms.update(ant.name() for ant in lemma.antonyms())
```

```
# Prints the antonyms if found.
if antonyms:
    print(f"Antonyms for '{token}': {', '.join(antonyms)}")
```

```
Antonyms Detection:
Antonyms for 'lead': follow, deficit
Antonyms for 'new': worn, old
Antonyms for 'has': refuse, lack, abstain
Antonyms for 'made': unmake, unmade, break
Antonyms for 'is': differ
Antonyms for 'looking': back
Antonyms for 'potential': actual
Antonyms for 'findings': lose
Antonyms for 'on': off
Antonyms for 'new': worn, old
Antonyms for 'soil': clean
Antonyms for 'local': express, national, general
Antonyms for 'hit': miss
Antonyms for 'double': multivalent, single, univalent
Antonyms for 'last': first
Antonyms for 'night': day
Antonyms for 'lead': follow, deficit
Antonyms for 'is': differ
Antonyms for 'respected': disesteem, disrespect
```

Homophones Detection (Alternate Approach)

```
print("\nHomophones Detection:")
# Homophones are generally detected using phonetic algorithms
or external libraries like `fuzzy` or `PyDictionary`.
# Here we might use a simple custom method based on pronunciation
or external data sources.
```

```
# Defines a function to check if two words are homophones based
on a simplified rule.
def is_homophone(word1, word2):
    return word1.lower() == word2.lower() and word1 != word2

# Example list of homophone pairs.
homophones_list = [('lead', 'led'), ('bare', 'bear'), ('pair',
'pear')]

# Iterates over each token.
for token in tokens:
    # Finds homophones for the current token based on the
      example list.
    homophones = [pair[1] for pair in homophones_list if
pair[0] == token.lower()]

    # Prints homophones if found.
    if homophones:
        print(f"Homophones for '{token}': {',
         '.join(homophones)}")
```

Homophones Detection:
Homophones for 'lead': led
Homophones for 'lead': led

Homographs Detection

```
# Prints the header for homograph detection.
print("\nHomographs Detection:")

# Iterates over each token.
for token in tokens:
    # Retrieves WordNet synsets for the token.
    synsets = wn.synsets(token)
```

```python
# Initializes a set to collect part-of-speech tags.
pos_tags = set()

# Adds POS tags for each synset to the set.
for synset in synsets:
    pos_tags.add(synset.pos())

# Prints if the token has multiple POS tags, indicating it
# is a homograph.
if len(pos_tags) > 1:
    print(f"'{token}' is a homograph with POS tags: {', '.join(pos_tags)}")
```

Homographs Detection:
'lead' is a homograph with POS tags: v, n
'engineer' is a homograph with POS tags: v, n
'in' is a homograph with POS tags: r, s, n
'new' is a homograph with POS tags: s, a, r
'bridge' is a homograph with POS tags: v, n
'project' is a homograph with POS tags: v, n
'has' is a homograph with POS tags: v, n
'made' is a homograph with POS tags: v, s, a
'team' is a homograph with POS tags: v, n
'now' is a homograph with POS tags: r, n
'looking' is a homograph with POS tags: v, s, n
'potential' is a homograph with POS tags: s, a, n
'impacts' is a homograph with POS tags: v, n
'findings' is a homograph with POS tags: v, n
'on' is a homograph with POS tags: a, r
'new' is a homograph with POS tags: s, a, r
'soil' is a homograph with POS tags: v, n
'Meanwhile' is a homograph with POS tags: r, n
'local' is a homograph with POS tags: a, n

'cricket' is a homograph with POS tags: v, n
'hit' is a homograph with POS tags: v, n
'double' is a homograph with POS tags: a, r, n, s, v
'in' is a homograph with POS tags: r, s, n
'sports' is a homograph with POS tags: v, n
'last' is a homograph with POS tags: a, r, n, s, v
'lead' is a homograph with POS tags: v, n
'engineer' is a homograph with POS tags: v, n
'project' is a homograph with POS tags: v, n
'team' is a homograph with POS tags: v, n
'respected' is a homograph with POS tags: v, s

Polysemy Detection

```
# Prints the header for polysemy detection.
print("\nPolysemy Detection:")
```

```
# Iterates over each token.
for token in tokens:
    # Retrieves WordNet synsets for the token.
    synsets = wn.synsets(token)

    # Prints if the token has more than one synset, indicating
    multiple meanings.
    if len(synsets) > 1:
        print(f"'{token}' has multiple meanings.")
```

Polysemy Detection:
'lead' has multiple meanings.
'engineer' has multiple meanings.
'in' has multiple meanings.
'new' has multiple meanings.
'bridge' has multiple meanings.

CHAPTER 6 SYNTACTIC AND SEMANTIC TECHNIQUES IN NLP

'project' has multiple meanings.
'has' has multiple meanings.
'made' has multiple meanings.
'a' has multiple meanings.
'breakthrough' has multiple meanings.
'team' has multiple meanings.
'is' has multiple meanings.
'now' has multiple meanings.
'looking' has multiple meanings.
'at' has multiple meanings.
'potential' has multiple meanings.
'impacts' has multiple meanings.
'findings' has multiple meanings.
'on' has multiple meanings.
'new' has multiple meanings.
'soil' has multiple meanings.
'Meanwhile' has multiple meanings.
'a' has multiple meanings.
'local' has multiple meanings.
'cricket' has multiple meanings.
'player' has multiple meanings.
'hit' has multiple meanings.
'double' has multiple meanings.
'ton' has multiple meanings.
'in' has multiple meanings.
'sports' has multiple meanings.
'news' has multiple meanings.
'last' has multiple meanings.
'night' has multiple meanings.
'lead' has multiple meanings.
'engineer' has multiple meanings.
'project' has multiple meanings.

'team' has multiple meanings.
'is' has multiple meanings.
'highly' has multiple meanings.
'respected' has multiple meanings.

Hyponyms of a Given Word

```
# Defines a function to find hyponyms for a given word.
def find_hyponyms(word):
    # Initializes a set to collect hyponyms.
    hyponyms = set()

    # Retrieves WordNet synsets for the given word.
    synsets = wn.synsets(word)

    # Iterates over each synset to find hyponyms.
    for synset in synsets:
        for hyponym in synset.hyponyms():
            for lemma in hyponym.lemmas():
                # Adds hyponyms to the set.
                hyponyms.add(lemma.name())

    # Returns the set of hyponyms.
    return hyponyms
```

Categorize the Article Based on Hyponyms of Given Categories

Listing 6-4 contains the code for this purpose

Listing 6-4. Categorize Articles

```
# Defines a function to categorize terms in the text based on
hyponyms of given categories.
def categorize_article(text, category_terms):
```

```
    # Tokenizes and lowercases the text.
    tokens = word_tokenize(text.lower())

    # Initializes a dictionary to store categorized terms.
    categorized_terms = {term: [] for term in category_terms}

    # Iterates over each token.
    for token in tokens:
        # Checks each category term for matching hyponyms.
        for term in category_terms:
            hyponyms = find_hyponyms(term)
            if token in hyponyms:
                # Appends the token to the corresponding
                category.
                categorized_terms[term].append(token)

    # Returns the dictionary of categorized terms.
    return categorized_terms

# Categorizes terms in the article text based on specified
general categories.
categories = categorize_article(article_text, ["scientist",
"research", "sports", "team", "player"])

# Print out the categorized terms or a message if none
are found.
for category, items in categories.items():
    if items:
        print(f"Category '{category}': {', '.join(items)}")
    else:
        print(f"Category '{category}': No matching
        terms found")
Category 'scientist': No matching terms found
Category 'research': No matching terms found
```

```
Category 'sports': No matching terms found
Category 'team': No matching terms found
Category 'player': lead, lead
```

6.4.8 Integration into Text Processing Pipelines

Any number of NLP techniques can be combined into a cohesive and convenient workflow pipeline to handle complex tasks more efficiently. By modularizing multiple functions, such as POS tagging, NER, and polysemy detection, we can create a comprehensive system (pipeline) that helps process and analyze text in a structured manner.

Such modular workflow integration pipelines permit seamless data flow between multiple stages of text analysis. Such workflow pipelines ensure that each next text processing step is built upon the results of the previous ones. Professional NLP application developers often use such pipelines. Next, Listing 6-5 demonstrates a simple text processing pipeline in just three steps.

Demonstrating Text Preprocessing Pipelines Using POS Tagging, NER, and Polysemy

Listing 6-5. A Simple Text Processing Pipeline

```
import nltk
from nltk.corpus import wordnet as wn
from nltk.tokenize import word_tokenize
from nltk import pos_tag, ne_chunk

# Ensure necessary nltk resources are downloaded.
nltk.download('punkt')
nltk.download('averaged_perceptron_tagger')
nltk.download('maxent_ne_chunker')
```

CHAPTER 6 SYNTACTIC AND SEMANTIC TECHNIQUES IN NLP

```python
nltk.download('words')
nltk.download('wordnet')

# Function for POS tagging.
def pos_tagging(tokens):
    return pos_tag(tokens)

# Function for NER.
def named_entity_recognition(pos_tags):
    return ne_chunk(pos_tags)

# Function for polysemy.
def find_polysemy(word):
    synsets = wn.synsets(word)
    return len(synsets) > 1

# Function to run the complete text processing pipeline.
def text_processing_pipeline(text):
    # Step 1: Tokenize the text
    tokens = word_tokenize(text)

    # Step 2: POS tagging
    pos_tags = pos_tagging(tokens)
    print("POS Tagging:")
    for word, tag in pos_tags:
        print(f"{word}: {tag}")

    # Step 3: Named Entity Recognition
    ner_tree = named_entity_recognition(pos_tags)
    print("\nNamed Entity Recognition:")
    for subtree in ner_tree:
        if hasattr(subtree, 'label'):
            print(f"{' '.join(c[0] for c in subtree)}: {subtree.label()}")
```

CHAPTER 6 SYNTACTIC AND SEMANTIC TECHNIQUES IN NLP

```
# Step 4: Polysemy detection
print("\nPolysemy Detection:")
for token in tokens:
    if find_polysemy(token):
        print(f"'{token}' has multiple meanings.")

# Sample article text.
article_text = """
The lead engineer for the new bridge project made a break-
through in New York. The team is now looking at the potential
impacts of their findings on the new soil. Meanwhile, Virat
Kohli hit his double ton in sports news last night. The lead
engineer of the project team is highly respected.
"""
# Run the complete text processing pipeline.
text_processing_pipeline(article_text)
OUTPUT:
POS Tagging:
The: DT
lead: NN
engineer: NN
for: IN
the: DT
new: JJ
bridge: NN
project: NN
made: VBD
a: DT
breakthrough: NN
in: IN
```

CHAPTER 6   SYNTACTIC AND SEMANTIC TECHNIQUES IN NLP

New: NNP
York: NNP
.: .
The: DT
team: NN
is: VBZ
now: RB
looking: VBG
at: IN
the: DT
potential: JJ
impacts: NNS
of: IN
their: PRP$
findings: NNS
on: IN
the: DT
new: JJ
soil: NN
.: .
Meanwhile: RB
,: ,
Virat: NNP
Kohli: NNP
hit: VBD
his: PRP$
double: JJ
ton: NN
in: IN
sports: NNS
news: NN

last: JJ
night: NN
.: .
The: DT
lead: JJ
engineer: NN
of: IN
the: DT
project: NN
team: NN
is: VBZ
highly: RB
respected: VBN
.: .

Named Entity Recognition:
New York: GPE
Virat Kohli: PERSON

Polysemy detection:
'lead' has multiple meanings.
'engineer' has multiple meanings.
'new' has multiple meanings.
'bridge' has multiple meanings.
'project' has multiple meanings.
'made' has multiple meanings.
'a' has multiple meanings.
'breakthrough' has multiple meanings.
'in' has multiple meanings.
'New' has multiple meanings.
'team' has multiple meanings.
'is' has multiple meanings.

'now' has multiple meanings.
'looking' has multiple meanings.
'at' has multiple meanings.
'potential' has multiple meanings.
'impacts' has multiple meanings.
'findings' has multiple meanings.
'on' has multiple meanings.
'new' has multiple meanings.
'soil' has multiple meanings.
'Meanwhile' has multiple meanings.
'hit' has multiple meanings.
'double' has multiple meanings.
'ton' has multiple meanings.
'in' has multiple meanings.
'sports' has multiple meanings.
'news' has multiple meanings.
'last' has multiple meanings.
'night' has multiple meanings.
'lead' has multiple meanings.
'engineer' has multiple meanings.
'project' has multiple meanings.
'team' has multiple meanings.
'is' has multiple meanings.
'highly' has multiple meanings.
'respected' has multiple meanings.

6.5 Chapter Recap

This chapter emphasizes the importance of understanding the structure and meaning of language, a key feature for the effective processing of human language.

CHAPTER 6   SYNTACTIC AND SEMANTIC TECHNIQUES IN NLP

The classification of words based on grammar, such as nouns or verbs, can be achieved through techniques like POS tagging. For this, Python tools such as NLTK and spaCy are utilized. Various parsing techniques, such as dependency parsing and constituency parsing, are explored in this chapter. Dependency parsing determines the relationship between words, whereas constituency parsing breaks the sentence into component phrases. Practical examples and challenges in syntactic analysis are presented to deepen understanding.

The latter part of the chapter dealt with semantics analysis and various methods to extract contextual meaning from text. One such technique is chunking, which is a method to group related words into meaningful units. The concepts of NER are presented for identifying names, places, and other specific entities in text. For words with multiple definitions, WSD helps in understanding their correct meaning. The importance of various matrix-based methods, such as term-document matrixes and TF–IDF, are also described to show their role in measuring word importance and relevance. Numerous coding tutorials are also included in this chapter to showcase the actual implementation of these techniques.

Concepts of lexical semantics are also discussed to examine the word relationships, such as synonyms, antonyms, homophones, homographs, polysemy, and hyponyms. It is very important to detect these relationships along with the integration of them into text processing pipelines.

A combination of syntactic and semantic techniques is indispensable for solving real-world problems concerning the structure and meaning of language. A classic case is **machine translation**, in which the correctness of syntax is crucial for grammatically accurate sentences in the target language. Semantics, on the other hand, assurances that the meaning of the original text is conserved.

The primary aim of studying these concepts is to infer and generate meaningful text from machine learning algorithms. This chapter concluded with some hands-on tutorials. The next chapter explains pragmatic analysis in NLP. Stay tuned!

6.6 References

[1] Coats, S. (2016). Grammatical feature frequencies of English on Twitter in Finland. English in computer-mediated communication: Variation, representation, and change, 93, 179-209.

[2] Pramod, Om (2023), Understanding POS Tagging: An In-Depth Exploration, Available at: `https://blog.gopenai.com/understanding-pos-tagging-an-in-depth-exploration-747f981d3514`, Last Accessed: 25th August 2024.

CHAPTER 7

Advanced Pragmatic Techniques and Specialized Topics in NLP

7.1 Why You Should Read This Chapter

Pragmatic analysis in NLP delves into understanding language in context. It goes beyond the literal meanings of words to interpret the intended message. Pragmatic analysis aims to discern the correct speaker's intent and the social or situational context from the input speech or text, taking into account factors such as speaker intent, tone, and the listener's perspective. Pragmatic analysis is essential for almost every serious NLP task, including more common ones like sentiment analysis, chatbot development, and conversational AI. In short, pragmatic analysis is responsible for the accurate interpretation of spoken or written text while considering the contexts of both leister and speaker.

Pragmatic analysis enables NLP applications to grasp the implied meanings, sarcasm, and indirect requests that are not directly conveyed

CHAPTER 7   ADVANCED PRAGMATIC TECHNIQUES AND SPECIALIZED TOPICS IN NLP

by the literal meanings of words but are understood through context. NLP applications integrated with pragmatic analysis are closely able to mimic human understanding. It leads to more natural and context-aware interactions between humans and machines (computers). Pragmatic analysis plays a pivotal role in NLP tasks dealing with dialogue systems and machine translation. Without learning pragmatic analysis, your aim of gaining any serious NLP skills is far from complete.

So far, you have learned about lexical, syntactic analysis, and semantic analysis. How do they relate to pragmatic analysis?

Lexical, syntactic, and semantic analyses are foundational steps that feed into pragmatic analysis in NLP. Lexical analysis is concerned with breaking down the text into words or tokens. After that, the syntactic analysis step categorizes these tokens into grammatical structures to enable grammatically correct sentence formation. Semantic analysis interprets only the literal meaning of these sentence structures while concentrating on the relationships between words. Pragmatic analysis is the final step, which incorporates input from the previous three layers and integrates context, speaker intent, and situational factors to accurately interpret the intended meaning of the text. Together, these four steps form a hierarchy, as shown in Figure 7-1. This chapter touches on disclosure integration along with the nuances of pragmatic analysis.

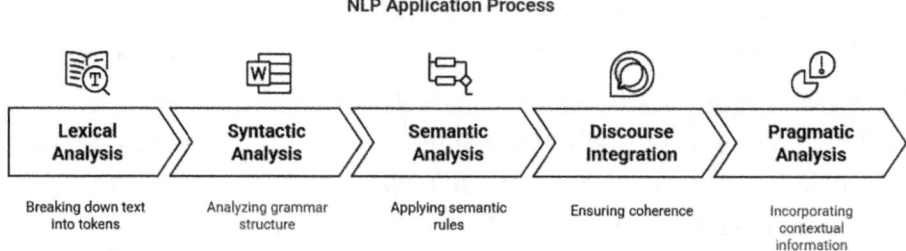

Figure 7-1. *The general process of most NLP applications*

7.2 Overview of Pragmatic Analysis Techniques

Let's discuss pragmatic analysis key concepts. We begin with distributional semantics and word embeddings, which are useful in understanding how words interact with one another in dynamic contexts. We study latent semantic analysis along with its real-world applications and practical applications to help us in the extraction of meaningful word relationships from large datasets. Word embeddings, such as Word2Vec, GloVe, CBOW, and skip-gram, are fundamental in representing words in continuous vector spaces. BERT and GPT further enhance word embeddings by integrating the surrounding context. It allows NLP algorithms to capture delicate shifts in word meanings. Topic modeling and techniques like Latent Dirichlet allocation help identify underlying themes in the text, which is pivotal to pragmatic analysis, especially while dealing with large corpora.

Finally, techniques such as ensemble methods and word similarity help us further improve the effectiveness of pragmatic analysis. They combine different models and measure contextual word relationships. Toward the end of this section, we introduce coreference resolution and its integration into NLP pipelines to further establish the importance of recognizing references.

7.3 Discourse Integration in NLP

In discourse integration, we keep track of an ongoing conversation to ensure that all the sentences fit together properly such that, as a collection, they make sense. In discourse integration, the focus is not on understanding just one sentence at a time. Before trying to make any sense, we integrate all the sentences and look at how each sentence connects with others. After that, we try to ensure that, together as a set, they make sense.

CHAPTER 7   ADVANCED PRAGMATIC TECHNIQUES AND SPECIALIZED TOPICS IN NLP

Let's try to solidify this concept with an example of a virtual assistant (e.g., Amazon's Alexa). You instruct your virtual assistant, "Turn on the TV," and then say, "Also, make it louder." The second sentence doesn't explicitly use the word "TV," but the assistant still analyzes what you are talking about. This is discourse integration; we are trying to make sense of this short conversation in totality. Without this, the assistant might get confused as we didn't mention "TV" in the second sentence.

In real life, we use discourse integration all the time. If your friend says, "I'm hungry," and then, "Let's get pizza," you know both sentences are related. In a similar fashion, discourse integration techniques help machines to follow and respond to conversations, just like humans. We often use pronouns (he, she, it) in our regular conversations. Discourse integration techniques used in NLP applications help to resolve how these pronouns are linked to the correct noun previously mentioned in the discourse. It also helps to interpret implied meanings and how context dynamically changes during a conversation. In another example, in automated customer service applications (chatbots), discourse integration helps the application to track the user's queries even over long conversations and respond in a relevant manner.

7.3.1 Techniques for Discourse Integration

The key techniques under discourse integration include coreference resolution, anaphora resolution, coherence modeling, rhetorical structure theory, discourse parsing, entity tracking, lexical chains, topic modeling, cohesion analysis, and temporal relation identification.

Coreference resolution enables NLP applications to connect the dots between words and their referents. For example, it ensures that "he," "she," or "it" connect to the right person or thing. For instance, in a news article, if it says, "Mary entered the class. She sat down," coreference resolution's job is to link "She" back to Mary. Anaphora resolution, on the other hand, zooms in on backward references. It resolves phrases like "this" or "that" to

CHAPTER 7 ADVANCED PRAGMATIC TECHNIQUES AND SPECIALIZED TOPICS IN NLP

their earlier mentions. For example, in the sentence, "I bought a new car. This is my favorite machine," anaphora resolution links "this" to "car."

Coherence modeling is like an invisible thread that helps keep a conversation or story connected and flowing smoothly. Imagine you're reading a book, and one chapter describes a character embarking on an adventure. Then, in the next chapter, it addresses the challenges the character faces. Coherence modeling helps link those chapters together so you can follow the journey without getting lost. For example, in a chatbot, if you ask, "What's the weather today?" and then, "Should I bring an umbrella?" coherence modeling helps the system understand that both questions are related to the weather.

Let's discuss the next technique with an example. In an essay, the opening paragraph describes the dangers of climate change and further explains possible solutions. Rhetorical structure theory (RST) reveals that the second part describes the first. RST delves to explain to us how different parts of a conversation or text are related. The RST role comes into play in explaining language constructs such as cause and effect, contrast, or explanation.

Let's study the case of a conversation between two friends. The first person might ask a question, and the other gives an answer. Discourse parsing is the technique that comes into play in separating these into "question" and "answer" units. Discourse parsing is equivalent to generating a blueprint of a conversation or text. It breaks the conversation or input text down into smaller, meaningful units to see how they work together. There are a couple of additional techniques related to discourse integration that are useful if you are aware of them. They are briefly explained in this chapter, and we link each technique with a real-world example for better understanding.

Entity tracking deals with keeping track of nouns, like people or things, throughout a story. For example, in a suspense story, it helps if you properly remember all the suspects and their actions that were mentioned in previous chapters. So, if the first chapter discusses a suspect named

John and the other chapter talks about Alex's friend, Mary, in this example, the job of entity tracking is to connect these details throughout the story so that machines have a clear understanding of the narrative.

Let's examine a different use case. You are reading a story about a family dog. Lexical chain techniques are used to connect all the related words like "dog," "puppy," and "pet" that may have mentioned different parts of the book. Keeping this link helps you to follow along without getting confused about what's being talked about. Lexical chains link connected words throughout the speech or input text to make sure everything sticks to the same topic. We use another example to discuss the topic modeling technique. Suppose you're reading a few news articles online. Topic modeling can group all the stories about climate change together. This can help you to easily find and read all the articles about the same topic.

In another instance, if an article starts by discussing a problem and then transitions to a solution, cohesion analysis is the technique that helps the machine ensure the changeover is clear and logical. Finally, Temporal relation identification helps to arrange events (in speech or text) in the order they occurred. In a history book, it helps to place events like wars and revolutions in chronological order. For example, it would indicate that World War I preceded World War II.

We discussed quite a few techniques in this section. We may not detail each one of them due to space constraints.

7.3.2 Demonstrating Techniques for Discourse Integration

Let's go over all the techniques and use Python code to demonstrate these processes.

CHAPTER 7   ADVANCED PRAGMATIC TECHNIQUES AND SPECIALIZED TOPICS IN NLP

Coreference Resolution (Code Demo)

Most of these techniques use popular NLP libraries in Python, such as spaCy, NLTK, Transformers, and Gensim, among others.

Here's a brief idea of how each technique can be demonstrated with code.

```
# Install the package needed for coreference resolution.
# Uncomment it if coreferee is not installed on your system.
!python -m pip install coreferee
# Installs the English coreference resolution model for the coreferee package.
!python -m coreferee install en
# Download the large English model ('en_core_web_lg') for spaCy.
# This model includes word vectors and is more accurate for NER and POS tagging.
# It's larger and more powerful than the smaller models lke 'en_core_web_sm'), but it may take up large memory.
# We have not loaded it in the main code body but without this step, the code was throwing error.
# So it's a required step.
!python -m spacy download en_core_web_lg
# Code to ignore warnings.
import warnings
warnings.filterwarnings("ignore", category=FutureWarning)
# Import the required package.
import spacy
# Load the transformer-based English model.
# As of now let's use it. We will take it up in detail in the later chapters.
nlp = spacy.load('en_core_web_trf')
```

CHAPTER 7   ADVANCED PRAGMATIC TECHNIQUES AND SPECIALIZED TOPICS IN NLP

```
"""
Add 'coreferee' to the pipeline.
'coreferee' is a spaCy extension that enables the
identification and linking of coreferences in a text.
Adding it to the pipeline so that coreference resolution can be
performed after other NLP tasks like tokenization and parsing.
"""
nlp.add_pipe('coreferee')
# Create the sample text for demo.
sample_doc = nlp(""" Although she was very busy at office work,
Mary felt she had had enough of it.
                    he and her spouse decided they needed to
                    go on a holiday.
                    They travelled by train to France
                    because they had enough friends in the
                    country.""")
print("")
print("OUTPUT \n")
sample_doc._.coref_chains.print()
```

OUTPUT

```
0: she(2), Mary(10), she(12), She(20), her(22)
1: work(8), it(17)
2: [She(20); spouse(23)], they(25), They(34), they(41)
3: France(39), country(47)
```

The output does not appear to be easy to understand. The first line of the production (indexed 0) tells us that the pronouns she(2), she(12), She(20), and her(22) refer to the same name, 'Mary(10).' Similarly, the second line (indexed 1) makes us understand that it(17) stands for work(8). At index 2 position, they(25), They(34), they(41) refer to [She(20);

282

spouse(23)]. Finally, at index 3, country(47) stands for France(39). Notice that in this output, all the pronouns in the sample_doc are resolved to their proper nouns (coreference resolution process).

Rhetorical Structure Theory (Code Demo)

```
!pip install stanza
import stanza

# Download and set up the Stanford NLP mode.
stanza.download('en')
{"model_id":"b1022cc610b04d679f12c5211b381398","version_major":2,"version_minor":0}
2024-09-08 16:28:37 INFO: Downloaded file to C:\Users\Shailendra Kadre\stanza_resources\resources.json
2024-09-08 16:28:37 INFO: Downloading default packages for language: en (English) ...
2024-09-08 16:28:38 INFO: File exists: C:\Users\Shailendra Kadre\stanza_resources\en\default.zip
2024-09-08 16:28:40 INFO: Finished downloading models and saved to C:\Users\Shailendra Kadre\stanza_resources
# Initialize a Stanza Pipeline for processing English text.
nlp = stanza.Pipeline(lang='en', processors='tokenize,mwt,pos, depparse, lemma')
2024-09-08 16:30:35 INFO: Checking for updates to resources.json in case models have been updated.  Note: this behavior can be turned off with download_method=None or download_method=DownloadMethod.REUSE_RESOURCES
{"model_id":"aaa69eecc06847fc9c18c0cbd0e1c88f","version_major":2,"version_minor":0}
2024-09-08 16:30:36 INFO: Downloaded file to C:\Users\Shailendra Kadre\stanza_resources\resources.json
```

CHAPTER 7 ADVANCED PRAGMATIC TECHNIQUES AND SPECIALIZED TOPICS IN NLP

```
2024-09-08 16:30:36 INFO: Loading these models for language: en (English):
=================================
| Processor | Package           |
---------------------------------
| tokenize  | combined          |
| mwt       | combined          |
| pos       | combined_charlm   |
| lemma     | combined_nocharlm |
| depparse  | combined_charlm   |
=================================

2024-09-08 16:30:36 INFO: Using device: cpu
2024-09-08 16:30:36 INFO: Loading: tokenize
2024-09-08 16:30:36 INFO: Loading: mwt
2024-09-08 16:30:36 INFO: Loading: pos
2024-09-08 16:30:36 INFO: Loading: lemma
2024-09-08 16:30:36 INFO: Loading: depparse
2024-09-08 16:30:37 INFO: Done loading processors!
# Process a text
doc = nlp("Your example sentence goes here.")
for sentence in doc.sentences:
    print("Tokens:", [word.text for word in sentence.words])
    print("POS Tags:", [word.pos for word in sentence.words])
    print("Dependencies:", [(word.text, word.deprel) for word
    in sentence.words])
Tokens: ['Your', 'example', 'sentence', 'goes', 'here', '.']
POS Tags: ['PRON', 'NOUN', 'NOUN', 'VERB', 'ADV', 'PUNCT']
Dependencies: [('Your', 'nmod:poss'), ('example', 'compound'),
('sentence', 'nsubj'), ('goes', 'root'), ('here', 'advmod'),
('.', 'punct')]
# Create the sample text for demo.
```

CHAPTER 7   ADVANCED PRAGMATIC TECHNIQUES AND SPECIALIZED TOPICS IN NLP

```
sample_doc = nlp(""" Although she was very busy at office work,
Mary felt she had had enough of it.
                    She and her spouse decided they needed to
                    go on a holiday.
                    They travelled by train to France
                    because they had enough friends in the
                    country.""")

# Process the text
doc = nlp(sample_doc)

# Print out tokens, POS tags, and dependency relations.
for sentence in doc.sentences:
    print("Sentence:", " ".join([word.text for word in
    sentence.words]))
    print("Tokens:", [word.text for word in sentence.words])
    print("POS Tags:", [word.pos for word in sentence.words])
    print("Dependencies:", [(word.text, word.deprel) for word
    in sentence.words])
    print("-----")
```
Sentence: Although she was very busy at office work , Mary felt she had had enough of it .
Tokens: ['Although', 'she', 'was', 'very', 'busy', 'at', 'office', 'work', ',', 'Mary', 'felt', 'she', 'had', 'had', 'enough', 'of', 'it', '.']
POS Tags: ['SCONJ', 'PRON', 'AUX', 'ADV', 'ADJ', 'ADP', 'NOUN', 'NOUN', 'PUNCT', 'PROPN', 'VERB', 'PRON', 'AUX', 'VERB', 'ADJ', 'ADP', 'PRON', 'PUNCT']
Dependencies: [('Although', 'mark'), ('she', 'nsubj'), ('was', 'cop'), ('very', 'advmod'), ('busy', 'advcl'), ('at', 'case'), ('office', 'compound'), ('work', 'obl'), (',', 'punct'),

CHAPTER 7   ADVANCED PRAGMATIC TECHNIQUES AND SPECIALIZED TOPICS IN NLP

('Mary', 'nsubj'), ('felt', 'root'), ('she', 'nsubj'), ('had', 'aux'), ('had', 'ccomp'), ('enough', 'obj'), ('of', 'case'), ('it', 'obl'), ('.', 'punct')]

Sentence: She and her spouse decided they needed to go on a holiday .
Tokens: ['She', 'and', 'her', 'spouse', 'decided', 'they', 'needed', 'to', 'go', 'on', 'a', 'holiday', '.']
POS Tags: ['PRON', 'CCONJ', 'PRON', 'NOUN', 'VERB', 'PRON', 'VERB', 'PART', 'VERB', 'ADP', 'DET', 'NOUN', 'PUNCT']
Dependencies: [('She', 'nsubj'), ('and', 'cc'), ('her', 'nmod:poss'), ('spouse', 'conj'), ('decided', 'root'), ('they', 'nsubj'), ('needed', 'ccomp'), ('to', 'mark'), ('go', 'xcomp'), ('on', 'case'), ('a', 'det'), ('holiday', 'obl'), ('.', 'punct')]

Sentence: They travelled by train to France because they had enough friends in the country .
Tokens: ['They', 'travelled', 'by', 'train', 'to', 'France', 'because', 'they', 'had', 'enough', 'friends', 'in', 'the', 'country', '.']
POS Tags: ['PRON', 'VERB', 'ADP', 'NOUN', 'ADP', 'PROPN', 'SCONJ', 'PRON', 'VERB', 'ADJ', 'NOUN', 'ADP', 'DET', 'NOUN', 'PUNCT']
Dependencies: [('They', 'nsubj'), ('travelled', 'root'), ('by', 'case'), ('train', 'obl'), ('to', 'case'), ('France', 'obl'), ('because', 'mark'), ('they', 'nsubj'), ('had', 'advcl'), ('enough', 'amod'), ('friends', 'obj'), ('in', 'case'), ('the', 'det'), ('country', 'nmod'), ('.', 'punct')]

CHAPTER 7 ADVANCED PRAGMATIC TECHNIQUES AND SPECIALIZED TOPICS IN NLP

We take the middle sentence for elaborations on the output.

- Sentence: "She and her spouse decided they needed to go on a holiday."

- Tokens and POS tags: ['She', 'and', 'her', 'spouse', 'decided', 'they', 'needed', 'to', 'go', 'on', 'a', 'holiday', '.']
POS Tags: ['PRON', 'CCONJ', 'PRON', 'NOUN', 'VERB', 'PRON', 'VERB', 'PART', 'VERB', 'ADP', 'DET', 'NOUN', 'PUNCT'] PRON (Pronoun): 'She', 'her', 'they' CCONJ (Coordinating Conjunction): 'and' NOUN (Noun): 'spouse', 'holiday' VERB (Verb): 'decided', 'needed, 'go' PART (Particle): 'to' ADP (Adposition): 'on' DET (Determiner): 'a' PUNCT (Punctuation): '.'

- Dependencies: ('She', 'nsubj'): 'She' is the nominal subject of the main verb 'decided'. ('and', 'cc'): 'and' is a coordinating conjunction linking 'She' and 'her spouse'. ('her', 'nmod:poss'): 'her' is a possessive modifier for 'spouse'. ('spouse', 'conj'): 'spouse' is a conjunct connected to 'She' by 'and'. ('decided', 'root'): 'decided' is the main verb (root) of the sentence. ('they', 'nsubj'): 'they' is the subject of the embedded verb 'needed'. ('needed', 'ccomp'): 'needed' is a clausal complement of 'decided'. ('to', 'mark'): 'to' is a marker for the infinitive verb 'go'. ('go', 'xcomp'): 'go' is an open clausal complement of 'needed'. ('on', 'case'): 'on' is a preposition marking the case of 'holiday'. ('a', 'det'): 'a' is a determiner for 'holiday'. ('holiday', 'obl'): 'holiday' is the oblique object of the preposition 'on'. ('.', 'punct'): '.' is punctuation marking the end of the sentence.

CHAPTER 7   ADVANCED PRAGMATIC TECHNIQUES AND SPECIALIZED TOPICS IN NLP

Having this information in your folds, you can now apply RST principles and manually complete the process of RST. More advanced tools are available to complete the RST process for you. Stanza provides valuable syntactic information. However, an end-to-end RST analysis would need additional steps or tools to explicitly categorize and recognize the rhetorical relationships between different text parts. To this date, a Python library for RST is unavailable to our knowledge.

Discourse Parsing (Code Demo)

A direct discourse parsing library is currently unavailable in Python. Professionals who write code often use advanced NLP libraries like spaCy or AllenNLP, along with deep learning techniques. We stick to spaCy for a simplified demo. The code is based on the following steps.

1. Load the spaCy model.

    – Initialize spaCy's English language model.

2. Define sample text.

    – Set a string variable with the sample text.

3. Process text.

    – Use spaCy to analyze the sample text.

    – Split the text into sentences.

4. Define a function to extract discourse information.

    – Do the following for each sentence in the processed text.

        a. Print the sentence.

        b. Extract and print named entities (if any).

c. Extract and print syntactic dependencies for each token.

d. Infer basic discourse relations based on keywords.

- If the sentence contains the word "because"
  - Record that this sentence provides a reason for the previous sentence.
- If the sentence contains the word "although"
  - Record that this sentence contrasts with the previous sentence.

5. Call a function to extract and display discourse information.

- Print inferred discourse relations.

```
!python -m spacy download en_core_web_sm
import spacy

# Load SpaCy's English model.
nlp = spacy.load('en_core_web_sm')

# Sample text.
text = """
Although she was very busy with office work, Mary felt she had had enough of it.
She and her spouse decided they needed to go on a holiday.
They travelled by train to France because they had enough friends in the country.
"""
```

CHAPTER 7 ADVANCED PRAGMATIC TECHNIQUES AND SPECIALIZED TOPICS IN NLP

```python
# Process the text with SpaCy.
doc = nlp(text)

# Function to extract and display discourse-like information
def extract_discourse_info(doc):
    sentences = list(doc.sents)
    relations = []

    for i, sent in enumerate(sentences):
        print(f"Sentence {i+1}: {sent.text}")

        # Extract named entities.
        entities = [(ent.text, ent.label_) for ent in
        sent.ents]
        if entities:
            print("  Named Entities:", entities)

        # Extract syntactic dependencies
        dependencies = [(token.text, token.dep_, token.head.
        text) for token in sent]
        print("  Dependencies:", dependencies)

        # Basic inference of discourse relations.
        if i > 0:
            previous_sent = sentences[i-1]
            if "because" in sent.text.lower():
                relations.append(f"Sentence {i+1} provides a
                reason for Sentence {i}")
            elif "although" in sent.text.lower():
                relations.append(f"Sentence {i+1} contrasts
                with Sentence {i}")

    return relations
```

CHAPTER 7   ADVANCED PRAGMATIC TECHNIQUES AND SPECIALIZED TOPICS IN NLP

 # Extract and print discourse information.
relations = extract_discourse_info(doc)
print("\nInferred Discourse Relations:")
for relation **in** relations:
    print(relation)
Sentence 1:

  Dependencies: [('\n', 'dep', '\n')]
Sentence 2: Although she was very busy with office work, Mary felt she had had enough of it.

  Named Entities: [('Mary', 'PERSON')]
  Dependencies: [('Although', 'mark', 'was'), ('she', 'nsubj', 'was'), ('was', 'advcl', 'felt'), ('very', 'advmod', 'busy'), ('busy', 'acomp', 'was'), ('with', 'prep', 'busy'), ('office', 'compound', 'work'), ('work', 'pobj', 'with'), (',', 'punct', 'felt'), ('Mary', 'nsubj', 'felt'), ('felt', 'ROOT', 'felt'), ('she', 'nsubj', 'had'), ('had', 'aux', 'had'), ('had', 'ccomp', 'felt'), ('enough', 'dobj', 'had'), ('of', 'prep', 'enough'), ('it', 'pobj', 'of'), ('.', 'punct', 'felt'), ('\n', 'dep', '.')]
Sentence 3: She and her spouse decided they needed to go on a holiday.

  Named Entities: [('a holiday', 'DATE')]
  Dependencies: [('She', 'nsubj', 'decided'), ('and', 'cc', 'She'), ('her', 'poss', 'spouse'), ('spouse', 'conj', 'She'), ('decided', 'ROOT', 'decided'), ('they', 'nsubj', 'needed'), ('needed', 'ccomp', 'decided'), ('to', 'aux', 'go'), ('go', 'xcomp', 'needed'), ('on', 'prep', 'go'), ('a', 'det', 'holiday'), ('holiday', 'pobj', 'on'), ('.', 'punct', 'decided'), ('\n', 'dep', '.')]

CHAPTER 7   ADVANCED PRAGMATIC TECHNIQUES AND SPECIALIZED TOPICS IN NLP

Sentence 4: They travelled by train to France because they had enough friends in the country.

```
Named Entities: [('France', 'GPE')]
Dependencies: [('They', 'nsubj', 'travelled'), ('travelled',
'ROOT', 'travelled'), ('by', 'prep', 'travelled'), ('train',
'pobj', 'by'), ('to', 'prep', 'travelled'), ('France',
'pobj', 'to'), ('because', 'mark', 'had'), ('they', 'nsubj',
'had'), ('had', 'advcl', 'travelled'), ('enough', 'amod',
'friends'), ('friends', 'dobj', 'had'), ('in', 'prep',
'friends'), ('the', 'det', 'country'), ('country', 'pobj',
'in'), ('.', 'punct', 'travelled'), ('\n', 'dep', '.')]
```

Inferred Discourse Relations:
Sentence 2 contrasts with Sentence 1
Sentence 4 provides a reason for Sentence 3

Discourse parsing deals with how different parts of a text are related to one another. It can be through logical and communicative connections. The main part of the output is the "discourse information."

The following input text is labeled by sentence numbers.

- (sentence 1) Although she was very busy with office work.

- (sentence 2) Mary felt she had had enough of it.

- (sentence 3) She and her spouse decided they needed to go on a holiday.

- (sentence 4) They traveled by train to France because they had enough friends in the country.

The following is the summary of discourse relations.

- **Contrasting relation**: Sentence 2 seems to contrast with sentence 1. The word "Although" in sentence 1 talks about a contrast with the decision pronounced in sentence 2.

- **Reason relation**: Sentence 4 seems to give a reason for the decision made in sentence 3. The word "because" specifies that sentence 4 explains why they traveled to France.

Entity Tracking (Code Demo)

Entity tracking deals with keeping track of the entities (like people, places, or objects) mentioned throughout a text. It's like keeping track of characters throughout a novel. Here's a simple code demo using spaCy for named entity recognition (NER) and tracking those entities.

```python
import spacy

# Load the English NLP model
nlp = spacy.load('en_core_web_sm')

# Sample text.
text = """
Although she was very busy with office work, Mary felt she had had enough of it.
She and her spouse decided they needed to go on a holiday.
They travelled by train to France because they had enough friends in the country.
"""
```

```python
# Process the text with spaCy
doc = nlp(text)

# Dictionary to track entities
entity_tracking = {}

# Iterate through the sentences in the doc
for sent in doc.sents:
    print(f"Sentence: {sent}")
    # Iterate through named entities in the sentence
    for ent in sent.ents:
        print(f"Entity: {ent.text}, Label: {ent.label_}")
        # Track entities and update if seen again
        if ent.text in entity_tracking:
            entity_tracking[ent.text] += 1
        else:
            entity_tracking[ent.text] = 1

# Output tracked entities
print("\nEntity Tracking:")
for entity, count in entity_tracking.items():
    print(f"{entity}: mentioned {count} times")
```
Sentence:

Sentence: Although she was very busy with office work, Mary felt she had had enough of it.

Entity: Mary, Label: PERSON
Sentence: She and her spouse decided they needed to go on a holiday.

Entity: a holiday, Label: DATE
Sentence: They travelled by train to France because they had enough friends in the country.

Entity: France, Label: GPE

Entity Tracking:
Mary: mentioned 1 times
a holiday: mentioned 1 times
France: mentioned 1 times

More professional code for entity tracking is written using packages like spaCy, AllenNLP, or Hugging Face Transformers. Entity code functions first detect, classify, and link entities across texts. These systems utilize NER and coreference resolution to track entities, even when they are pronouns or synonyms. Professional code is often part of larger NLP pipelines that integrate with databases or knowledge graphs to establish entity relationships and maintain the required accuracy across long input documents or conversations.

Lexical Chains (Code Demo)

A lexical chain is a sequence of related words that are connected either through direct synonyms or semantically related terms. These related words share a common meaning or topic. We discussed lexical chains with an example in the theory section. This book provides a simple code demo of lexical chains using WordNet from the NLTK library.

For professional NLP applications, lexical chain code utilizes advanced algorithms that may be based on WordNet, distributional semantics, or word embeddings, such as Word2Vec or BERT, to capture semantic relationships between words. Such systems use synonyms, hypernyms, and context-based similarities to generate accurate lexical chains. This code is often united with text segmentation, word sense disambiguation, and coherence modeling for tasks involving summarization or topic detection.

```python
# !pip install nltk
# !python -m nltk.downloader wordnet
import nltk
from nltk.corpus import wordnet as wn

# Sample text
text = "Anil is a blood student in my class. He runs very fast."

# Tokenize the text
words = nltk.word_tokenize(text.lower())

# Function to find synonyms from WordNet
def get_synonyms(word):
    synonyms = set()
    for syn in wn.synsets(word):
        for lemma in syn.lemmas():
            synonyms.add(lemma.name())
    return synonyms

# Build lexical chains
lexical_chains = []

for word in words:
    found_chain = False
    word_synonyms = get_synonyms(word)

    # Check if word fits into any existing chain
    for chain in lexical_chains:
        if chain.intersection(word_synonyms):
            chain.update(word_synonyms)
            found_chain = True
            break
```

CHAPTER 7   ADVANCED PRAGMATIC TECHNIQUES AND SPECIALIZED TOPICS IN NLP

```
# If not, start a new chain
if not found_chain and word_synonyms:
    lexical_chains.append(set(word_synonyms))

# Output lexical chains
for i, chain in enumerate(lexical_chains):
    print(f"Chain {i + 1}: {chain}")
```
Chain 1: {'indigo', 'indigotin', 'Indigofera_suffruticosa', 'anil', 'Indigofera_anil'}
Chain 2: {'personify', 'represent', 'be', 'equal', 'follow', 'live', 'cost', 'constitute', 'embody', 'comprise', 'make_up', 'exist'}
Chain 3: {'type_A', 'adenine', 'ampere', 'a', 'angstrom_unit', 'antiophthalmic_factor', 'vitamin_A', 'angstrom', 'axerophthol', 'amp', 'group_A', 'deoxyadenosine_monophosphate', 'A'}
Chain 4: {'debauched', 'lineage', 'line', 'riotous', 'blood', 'blood_line', 'fast', 'degenerate', 'quick', 'bloodline', 'dissolute', 'flying', 'libertine', 'tight', 'fasting', 'firm', 'parentage', 'rake', 'ancestry', 'degraded', 'loyal', 'profligate', 'rip', 'rakehell', 'stock', 'dissipated', 'immobile', 'stemma', 'pedigree', 'descent', 'truehearted', 'line_of_descent', 'roue', 'origin'}
Chain 5: {'scholarly_person', 'bookman', 'pupil', 'educatee', 'student', 'scholar'}
Chain 6: {'Indiana', 'atomic_number_49', 'In', 'IN', 'inward', 'Hoosier_State', 'inch', 'inwards', 'indium', 'in'}
Chain 7: {'execute', 'division', 'incline', 'family', 'outpouring', 'bleed', 'runnel', 'consort', 'running_play', 'bunk', 'pass', 'persist', 'social_class', 'be_given', 'campaign', 'run_for', 'escape', 'hunt_down', 'discharge', 'running_game', 'ply', 'lam', 'guide', 'trial', 'hightail_it',

CHAPTER 7 ADVANCED PRAGMATIC TECHNIQUES AND SPECIALIZED TOPICS IN NLP

'classify', 'category', 'go', 'lead', 'head_for_the_hills', 'rivulet', 'sort', 'rill', 'lean', 'carry', 'unravel', 'turn_tail', 'course_of_study', 'melt_down', 'move', 'streak', 'separate', 'foot_race', 'political_campaign', 'ravel', 'fly_the_coop', 'stratum', 'scat', 'race', 'running', 'course_of_instruction', 'melt', 'run_away', 'scarper', 'streamlet', 'function', 'work', 'endure', 'draw', 'hunt', 'test', 'black_market', 'tally', 'operate', 'sort_out', 'track_down', 'play', 'run', 'break_away', 'tend', 'ladder', 'form', 'grade', 'feed', 'die_hard', 'footrace', 'class', 'flow', 'assort', 'extend', 'take_to_the_woods', 'socio-economic_class', 'course', 'year', 'prevail', 'range'}
Chain 8: {'atomic_number_2', 'helium', 'He', 'he'}
Chain 9: {'very', 'real', 'really', 'selfsame', 'rattling', 'identical'}

The output characterizes lexical chains as groups of semantically related words from the input text. Sometimes, you see a couple of unrelated terms (not given in the input text) in these lexical chains. This is an example of how lexical chaining can bring in terms that aren't explicitly present but are linked conceptually in the word database. The algorithm sometimes needs fine-tuning to limit the chain generation to the terms that more closely match your input text.

Topic Modeling (Code Demo)

Topic modeling techniques use word patterns and groupings to identify the key themes or topics in a large collection of texts. This chapter presents a simple demo of topic modeling using the TF-IDF technique. (We covered this technique in earlier chapters.) Professional topic modeling code is often written using advanced techniques like Latent Dirichlet allocation

(LDA) or non-negative matrix factorization. It leverages Python packages like Gensim or scikit-learn. LDA is covered in the upcoming chapters.

We follow these five simple steps to write our code.

1. Prepare documents. Start with a list of text documents you want to analyze.

2. Initialize vectorizer. Create a TfidfVectorizer object. A TF–IDF score converts the input text documents into numerical data. This process ignores stop words.

3. Transform text. Use the vectorizer to convert the input text documents into a TF–IDF matrix.

4. Get feature names. Extract the list of words that were considered in the TF–IDF analysis.

5. Display scores. For each document, print the words and their TF–IDF scores.

```
from sklearn.feature_extraction.text import TfidfVectorizer

# Sample text data (documents)
documents = ["Although she was very busy with office work, Mary felt she had had enough of it.",
             "She and her spouse decided they needed to go on a holiday.",
             "They travelled by train to France because they had enough friends in the country."]

# Initialize the TF-IDF Vectorizer
tfidf_vectorizer = TfidfVectorizer(stop_words='english')

# Fit and transform the documents into a TF-IDF matrix
tfidf_matrix = tfidf_vectorizer.fit_transform(documents)
```

```python
# Get the feature names (terms) from the TF-IDF model
feature_names = tfidf_vectorizer.get_feature_names_out()

# Display the TF-IDF scores for each document
for doc_idx, doc in enumerate(documents):
    print(f"\nDocument {doc_idx + 1}: {doc}")
    # Get the TF-IDF scores for each word in the document
    for word_idx in tfidf_matrix[doc_idx].nonzero()[1]:
        print(f"{feature_names[word_idx]}: {tfidf_matrix[doc_idx, word_idx]:.4f}")
```

Document 1: Although she was very busy with office work, Mary felt she had had enough of it.
felt: 0.4472
mary: 0.4472
work: 0.4472
office: 0.4472
busy: 0.4472

Document 2: She and her spouse decided they needed to go on a holiday.
holiday: 0.5000
needed: 0.5000
decided: 0.5000
spouse: 0.5000

Document 3: They travelled by train to France because they had enough friends in the country.
country: 0.4472
friends: 0.4472
france: 0.4472
train: 0.4472
travelled: 0.4472

The output addresses the TF-IDF scores for the words in documents 1, 2, and 3. The higher the score, the more relevant the word is to the content of the document. In the case of document 2, each of these words has a high score of 0.5000. It means all the words are equally important in this document. If a word has a high TF-IDF score, it means that the word is exclusive to that document, and it could be a strong pointer of its topic or content. Note that this was a simplified analysis. Better results can be obtained using advanced techniques like LDA.

Cohesion Analysis (Code Demo)

Cohesion analysis has two objectives. First, it looks at how well different parts of a text fit together. And second, how they connect to make the text flow smoothly and make sense. The following demo brings in a new concept of cosine similarity, which measures the similarity between two vectors. An interpretation of the similarity scores is given. We cover this concept in later in this chapter.

We used a basic TF-IDF technique in our code demo. More advanced vectorization techniques used by professionals for cohesion insight include word embeddings, such as Word2Vec and GloVe, to capture more nuanced insights into text similarity and cohesion. We take up these techniques later in this chapter.

```
from sklearn.feature_extraction.text import TfidfVectorizer
from sklearn.metrics.pairwise import cosine_similarity

# Revised sentences
sentences = [
    "Mary is feeling overwhelmed with her busy office job but
    finds some relief in her evening walks.",
    "Mary feels overwhelmed with her hectic work schedule, yet
    she finds relaxation in her daily evening walks.",
```

    "Despite a busy workday, Mary enjoys unwinding with a walk
    in the evening."
]

```
# Initialize the TF-IDF Vectorizer
vectorizer = TfidfVectorizer(stop_words='english')

# Fit and transform the sentences into a TF-IDF matrix
tfidf_matrix = vectorizer.fit_transform(sentences)

# Compute the cosine similarity matrix
cosine_sim_matrix = cosine_similarity(tfidf_matrix)

# Display the cosine similarity matrix
print("Cosine Similarity Matrix:")
for i in range(len(sentences)):
    for j in range(len(sentences)):
        print(f"Similarity between Sentence {i + 1} and
        Sentence {j + 1}: {cosine_sim_matrix[i, j]:.2f}")
Cosine Similarity Matrix:
Similarity between Sentence 1 and Sentence 1: 1.00
Similarity between Sentence 1 and Sentence 2: 0.32
Similarity between Sentence 1 and Sentence 3: 0.19
Similarity between Sentence 2 and Sentence 1: 0.32
Similarity between Sentence 2 and Sentence 2: 1.00
Similarity between Sentence 2 and Sentence 3: 0.10
Similarity between Sentence 3 and Sentence 1: 0.19
Similarity between Sentence 3 and Sentence 2: 0.10
Similarity between Sentence 3 and Sentence 3: 1.00
```

Interpretation of the output consists of the following.

- **Interpreting the scores**: Higher scores specify more cohesive sentences with a shared vocabulary and similar content.

- **Low inter-sentence similarity**: The highest is 0.32. It shows limited shared content or vocabulary.

- **Minimal overlap**: Low similarity scores of 0.10 and 0.19 indicate minimal thematic or lexical overlap between the sentences.

A similarity score of 1.00 indicates perfect similarity with itself. Similarity scores of 0 indicate no shared vocabulary and similarity of content between sentences.

Temporal Relation Identification (Code Demo)

Temporal relation identification is the process of finding time-based relationships between events in an input text. For instance, take the sentence "I want to complete this chapter before lunch;" the word "before" indicates the time-based connection between two events. It describes finishing work earlier than lunch.

Our code demo (Listing 7-1) for temporal relation identification would first perform dependency parsing to extract events and detect temporal signals, which include words like "before," "after," "during," "until," and "while." Following this, the classification of events is performed using rule-based methods or machine learning models. This approach aims to find the sequence and timing of actions in a given input text. We utilize spaCy to extract events.

Our program follows these five steps.

1. Import libraries and load models.
2. Feature extraction from sentences.
3. Prepare training data.
4. Train random forest classifier.
5. Predict temporal relations in new sentences.

Listing 7-1. Temporal Relation Identification

```python
# Required Libraries
import spacy
from sklearn.ensemble import RandomForestClassifier
import numpy as np

# Load the spaCy model for dependency parsing
nlp = spacy.load("en_core_web_sm")

# Sample sentences with temporal relations
sentences = [
    "She finished her work before going to lunch.",
    "He went to the gym after work.",
    "They waited until the show started.",
    "The meeting was delayed during the storm."
]

# Temporal signals
temporal_signals = ["before", "after", "during", "until", "while"]

# Feature extraction - identifying events and temporal signals
def extract_features(sent):
    doc = nlp(sent)
    events = []
    temporal_relation = ""

    for token in doc:
        if token.dep_ == "ROOT":  # Event (verb) extraction
            events.append(token.lemma_)
        if token.text in temporal_signals:  # Temporal signal detection
            temporal_relation = token.text
    return events, temporal_relation
```

```python
# Prepare training data - sentences, events, and labels (1 for
temporal relation present, 0 otherwise)
X = []
y = []

for sentence in sentences:
    events, signal = extract_features(sentence)
    if signal:  # If temporal signal is present, we
    classify as 1
        X.append([len(events)])  # Simple feature: number
        of events
        y.append(1)
    else:
        X.append([0])
        y.append(0)

# Convert to numpy arrays
X = np.array(X)
y = np.array(y)

# Train a Random Forest Classifier
clf = RandomForestClassifier(n_estimators=10, random_state=42)
clf.fit(X, y)

# Predict on new sentences
new_sentences = ["John started his project before the deadline.",
                 "She arrived after the party had begun."]

for new_sent in new_sentences:
    events, signal = extract_features(new_sent)
    prediction = clf.predict([[len(events)]])
    print(f"Sentence: {new_sent}")
```

```
print(f"Detected Temporal Signal: {signal}, Prediction:
{'Temporal relation' if prediction == 1 else 'No temporal
relation'}\n")
```
```
Sentence: John started his project before the deadline.
Detected Temporal Signal: before, Prediction: Temporal relation

Sentence: She arrived after the party had begun.
Detected Temporal Signal: after, Prediction: Temporal relation
```

function for better understanding.

- **extract_features(sent)** processes each sentence using spaCy. The function returns the list of events (verbs) and the temporal relation signal (if present).

- **doc = nlp(sent)** prepares the sentence and puts it into a structured format for analysis.

- **for token in doc** loops through each word in the sentence.

- **token.dep_ == "ROOT"** checks if the word is the main verb (the root action) and stores it.

- **if token.text in temporal_signals**: If a word is found in the list of temporal signals, it is stored as a temporal signal.

Converting words into ML model format: Without vectorizing the entire sentence, the logic uses a simple logic for converting words into numbers. It counts the number of verbs (events) as features in X and puts a binary label (1 or 0) for y based on the presence of temporal signals. This keeps the dataset simple and in numerical format without using a vectorizer.

The output confirms that the code is effectively detecting temporal signals present in the input sentences. The model also correctly predicts the existence of temporal relations based on these signals.

7.4 Distributional Semantics and Word Embeddings

Distributional semantics deals with discovering word meanings based on the company they keep. Words appearing in similar contexts often have interrelated meanings. Let's look at an example. "Doctor" and "nurse" are often used together in hospital and healthcare documents; we can assume they are related in meaning. Let's look at another example: the words "king" and "queen" are often found together in contexts related to royalty, palaces, or authority. This way, they are seen as similar. Distributional semantics works based on the **distributional hypothesis,** which states, "You shall know a word by the company it keeps."

To recognize word meanings, each word is assigned a unique vector, which appears as a list of numbers. These **word vectors** are created using a large amount of text data. If two words have meanings, the vectors for both are close to each other. Similar words have less Euclidean distance or cosine similarity (discussed later) between the vectors. With these numerical forms of words, we can measure similarity. These word vectors are sometimes used interchangeably in meaning with "word embeddings."

Let's look at word vectors (word embeddings). The actual numerical values of word embeddings for words like "king," "man," "woman," and "queen" can change depending on the type of word embedding model used. There are many word embedding models in use; a few examples are Word2Vec, GloVe, and FastText. To give you an idea, we use the hypothetical word embeddings of king, man, and woman as follows.

- king: 0.6,0.4,0.1,−0.2,0.3,0.5,−0.10
- man: 0.3,0.5,0.2,−0.1,0.1,0.4,0.00
- woman: 0.4,0.3,0.5,0.0,0.2,0.6,0.1

If you do a simple math (king – man) + woman, you get the calculated value of queen.

- Queen$_c$: 0.7,0.2,0.4, −0.1,0.4,0.7,0.0

    The calculated vector Queen$_c$ is close to the following hypothetical vector of queen (Queen$_h$).

- **Queen$_h$:** 0.7,0.5,0.3,0.1,0.4,0.8,0.2

Here, we used the classic example of queen ≈ king – man + woman, which demonstrates how word embeddings can capture relationships between words through mathematical operations. Word embeddings are useful for various NLP tasks like machine translation and sentiment analysis.

Euclidean distance is the distance between two-word vectors in space. Let's look at a simple example of two-word vectors with hypothetical numerical values as follows. To keep it simple, we take only three numbers in each word vector.

- king: [0.8, 0.5, 0.6]
- queen: [0.7, 0.6, 0.5]
- apple: [0.2, 0.1, 0.3]

The Euclidean distance "king" and "queen" is calculated as follows.

Euclidean distance (king, queen) = SQUARE ROOT $((0.8-0.7)^2+(0.5-0.6)^2+(0.6-0.5)^2) = 0.173$

Euclidean distance (king, apple) = SQUARE ROOT $((0.8-0.2)^2+(0.5-0.1)^2+(0.6-0.3)^2) = 0.781$

As expected, Euclidean distance (king, queen) is lesser than Euclidean distance (king, apple). The reason is simple: *king* and *queen* are similar in meaning because both belong to a royal family, but *king* and *apple* have different meanings.

CHAPTER 7 ADVANCED PRAGMATIC TECHNIQUES AND SPECIALIZED TOPICS IN NLP

Similar to Euclidean distance, cosine similarity also measures how close or similar word vectors are. Euclidean distance is based on calculating the actual distance between vectors, while cosine similarity measures the angle between them. **Cosine similarity** focuses on direction rather than magnitude. Let's calculate the cosine similarity between two words through an example. The following uses the same word vectors of "king", "queen", and "apple" with the same hypothetical numerical values.

The following steps calculate the cosine similarity between the words "king" and "queen".

- Dot product of "king" and "queen" word vectors = (0.8×0.7)+(0.5×0.6)+(0.6×0.5)=1.16

- Magnitude of "kind" word vector = SQUARE ROOT($(0.8)^2+(0.5)^2+(0.6)^2$)=1.118

- Magnitude of "queen" word vector= SQUARE ROOT($(0.7)^2+(0.6)^2+(0.5)^2$)= 1.049

- Cosine similarity = 1.16/ (1.118×1.049) = 0.989

Similarly, if we compute the cosine similarity between "king" and "apple," it comes out as 0.933. Note that the cosine similarity of "king" and "queen" (0.989) is more than the cosine similarity of "king" and "apple" (0.933). It indicates a closer relationship between the former pair compared to the latter. Cosine similarity ranges from −1 to 1.

- 1 indicates perfect similarity (parallel vectors that point in the same direction).

- 0 means no similarity (vectors are orthogonal).

- −1 indicates perfect dissimilarity (vectors point in opposite directions).

CHAPTER 7   ADVANCED PRAGMATIC TECHNIQUES AND SPECIALIZED TOPICS IN NLP

Cosine similarity is often used in information retrieval to assess the degree to which two documents or words are related in meaning. Euclidean distance is frequently used in NLP tasks such as document clustering.

7.4.1 Latent Semantic Analysis

Word embeddings enable machines to process natural languages word by word. They convert words into numerical representations, which machines (often computers) understand. This way, machines can interpret and process language more effectively. Over the years, several models have emerged to develop word embeddings, including Word2Vec and GloVe, as well as contextualized embeddings such as BERT and GPT. These approaches help model the delicate nuances of language, enabling NLP applications to identify semantic similarity, context, and deeper language structures. This section explores these techniques, from traditional embeddings to the latest advancements, to understand how they contribute to language understanding in NLP applications. This section discusses latent semantic analysis (LSA), which is a foundational technique to develop word embeddings.

LSA is a technique used to determine the meaning of words based on their context. It examines large text corpora to identify patterns showing how words relate to each other. Here, *patterns* refer to the ways words are frequently found together in similar contexts. This frequency reveals that words often convey related ideas or themes when they appear in similar contexts.

We can use either the term-document matrix (TDM) or the TF-IDF matrix for LSA, but they serve different purposes. TF-IDF is preferred because it helps LSA focus on the most important words in documents. TF-IDF puts more weight on exceptional words, and it helps us realize the main ideas. (TDM and TF-IDF were covered earlier in this book.) The following steps preconform LDA using TF-IDF.

1. Clean and format the data for analysis.

2. Construct the TF–IDF matrix from the cleaned documents.

3. Apply singular value decomposition (SVD) to the TF–IDF matrix to reduce its dimensions and to identify latent topics within the text.

4. Analyze the results to figure out term-document relationships and identify prominent themes or topics.

5. Visualize the reduced matrix or the results of the LDA to better understand patterns and meanings in the data.

Code Demo of LSA

Listing 7-2 is a code demo of LSA using a simplified text corpus. We use the TF–IDF matrix in this demo.

Listing 7-2. LSA Code Demo

```
# The following is a simplified code demo of LSA
import pandas as pd
import numpy as np
import matplotlib.pyplot as plt
from sklearn.feature_extraction.text import TfidfVectorizer
from sklearn.decomposition import TruncatedSVD

# Sample common text corpus
documents = [
    "The cat sat on the mat.",
    "Dogs are great companions.",
    "The sun is bright today.",
```

CHAPTER 7   ADVANCED PRAGMATIC TECHNIQUES AND SPECIALIZED TOPICS IN NLP

```
    "Cats and dogs are popular pets.",
    "The weather is nice for a walk."
]

# Step 1: Data Cleaning and Formatting for Analysis,
# Skipping this step as the data has already been cleaned.

# Step 2: Construct the TF-IDF matrix
vectorizer = TfidfVectorizer()
tfidf_matrix = vectorizer.fit_transform(documents)

# Step 3: Apply SVD to reduce dimensions
svd = TruncatedSVD(n_components=2)
lsa_matrix = svd.fit_transform(tfidf_matrix)

# Step 4: Analyze the results
# Create a DataFrame for viewing
lsa_df = pd.DataFrame(lsa_matrix, columns=['Concept 1',
'Concept 2'])
print("Reduced Dimensions (Latent Concepts):")
print(lsa_df)
Reduced Dimensions (Latent Concepts):
      Concept 1      Concept 2
0   6.263182e-01   1.932910e-16
1  -4.140572e-16   8.096747e-01
2   7.173435e-01  -7.926662e-15
3  -1.891549e-16   8.096747e-01
4   6.985829e-01   8.989044e-15
```

Let's interpret this output matrix.

- In the output matrix, rows represent individual documents.

- Columns represent latent concepts or themes.

- Values indicate the strength of each document's relationship to those concepts.

- Negative values indicate a lack of relevance of that document to that concept (column).

- High values indicate more relevance of that document (row) to that concept (column).

- The first and third documents (column 1) show high values. It shows strong relevance to Concept 1.

- The second document has a significant value (column 2). It shows strong relevance to Concept 2.

- Other documents in both columns are less relevant.

- The first and third documents may share a theme.

- The second document relates to a different topic.

- Higher scores indicate thematic similarities among documents. It helps in clustering and topic identification.

- Look at which documents have higher values for each concept and then check their content.

- You can then decode what concept 1 and concept 2 represent based on their thematic similarities.

7.4.2 Popular Word Embeddings

Word embeddings use large amounts of text to learn the relationship between the words. Popular word embeddings, such as Word2Vec, GloVe, and FastText, capture context, similarity, and semantic nuances to map words into numerical vectors. These numerical vectors essentially capture their meaning and relationships between words. Next, let's divide these word embeddings into traditional and newer ones, like BERT and GPT, based on contextual word representations.

Traditional Word Embeddings

Traditional word embeddings map words to fixed-size vectors. They were the early techniques in NLP based on word co-occurrence patterns in large text datasets. Traditional techniques, such as Word2Vec and GloVe, analyze how frequently and in what context words appear together in large training datasets to capture relationships between words and create word representations in the form of word vectors.

Word2Vec

Word2Vec (word to vector) converts words to machine readable numeric vectors, which represent a specific word (as a vector) in multidimensional space. These vectors help machines to capture word meaning, semantic similarity, and relationship with surrounding text. Word2Vec is essentially a pre-trained model. It utilizes a shallow neural network model to acquire the meaning of words from a large corpus of (training) texts. To make the processing prompt, fast, and transparent, Word2Vec neural networks use only one or two hidden layers. These neural networks are trained using large databases of texts. The Word2Vec algorithm uses either continuous bag of words or skip-gram approaches to generate word embeddings.

Continuous Bag of Words Model

The continuous bag of words (CBOW) model is an unsupervised method of finding word embeddings. It predicts the target word by utilizing the words surrounding it (context words). For this purpose, it utilizes a shallow neural network program (CBOW model). This program learns to predict any target word by utilizing the words that appear before and after it in

CHAPTER 7 ADVANCED PRAGMATIC TECHNIQUES AND SPECIALIZED TOPICS IN NLP

a given context window. By concentrating on the surrounding context words, the CBOW model is able to capture the meaning of a word (in a given context) in the form of numerical vectors (embeddings).

For instance, in the sentence, "he is a great scholar," if we want the word embeddings of the word "great," we first take a context window. Assuming this context window has two words, then we consider two words before the target word "great" and also consider two words after it. So, for our example sentence, the context words are as follows.

- Two words before the target word "great" – ("is", "a")
- Two words after the target word "great" – ("scholar")

Using this context window, the CBOW model learns to calculate the word embeddings of the target word "great." Listing 7-3 is a code demonstration of CBOW, a well-documented shallow neural network.

Listing 7-3. Code Demo of Continuous Bag of Words Model

```
8 #Running a shallow neural network program to
demonstrate CBOW.
9 # Ignore deprecation warnings for cleaner output.
import warnings
warnings.filterwarnings("ignore", category=DeprecationWarning)

# Import libraries
import numpy as np
import tensorflow as tf
from tensorflow.keras.models import Sequential
# Sequential model
from tensorflow.keras.layers import Dense, Embedding,
Flatten  # Layers for neural network
from tensorflow.keras.preprocessing.text import Tokenizer
# Tokenizer for text preprocessing
```

CHAPTER 7   ADVANCED PRAGMATIC TECHNIQUES AND SPECIALIZED TOPICS IN NLP

```python
from tensorflow.keras.preprocessing.sequence import pad_
sequences  # For padding sequences

# Sample corpus of sentences (training curpus)
sentences = [
    "he is a great scholar",
    "a great scholar writes great papers",
    "the scholar is very great",
    "great ideas come from great minds"
]

# Tokenize the sentences
tokenizer = Tokenizer()
tokenizer.fit_on_texts(sentences)  # Fit tokenizer on sentences
total_words = len(tokenizer.word_index) + 1  # Total unique
words in corpus

# Initialize lists for CBOW input-output pairs
input_data = []   # Context words
output_data = []  # Target word

# Define the context window size
window_size = 2

# Create input-output pairs for CBOW
for sentence in sentences:
    words = sentence.split()
  # Loop through the words
    for i in range(window_size, len(words) - window_size):
        context = []
        for j in range(i - window_size, i + window_size + 1):
            if j != i:  # Skip target word
                context.append(tokenizer.word_
                index[words[j]])  # Append context word index
```

```python
        input_data.append(context)  # Add context to input data
        output_data.append(tokenizer.word_index[words[i]])
        # Add target word to output data

# Pad input sequences to ensure uniform length
input_data = pad_sequences(input_data, padding='post')

# One-hot encode output data for categorical labels
output_data = np.array(output_data)
output_data = np.eye(total_words)[output_data]  # Convert
to one-hot

# Build CBOW model
model = Sequential()
model.add(Embedding(input_dim=total_words, output_dim=10,
input_length=window_size * 2))  # Embedding layer
model.add(Flatten())  # Flatten to 1D
# Output layer with softmax for word prediction
model.add(Dense(total_words, activation='softmax'))

# Compile model with optimizer, loss, and metrics
model.compile(optimizer='adam', loss='categorical_
crossentropy', metrics=['accuracy'])

# Train the model on input-output pairs
model.fit(input_data, output_data, epochs=20, verbose=1)

# Retrieve embedding for the word "great"
# Index of "great" in vocabulary
great_index = tokenizer.word_index['great']
great_embedding = model.layers[0].get_weights()[0][great_
index]  # Extract embedding
print("Embedding for 'great':", great_embedding)
10 Epoch 1/20
```

CHAPTER 7   ADVANCED PRAGMATIC TECHNIQUES AND SPECIALIZED TOPICS IN NLP

```
11 1/1 ───────────────────────── 1s 725ms/step -
accuracy: 0.1667 - loss: 2.6119
Epoch 2/20
1/1 ────────────────────────── 0s 19ms/step -
accuracy: 0.3333 - loss: 2.6053
Epoch 3/20
1/1 ────────────────────────── 0s 21ms/step -
accuracy: 0.5000 - loss: 2.5986
.......
Epoch 20/20
1/1 ────────────────────────── 0s 25ms/step -
accuracy: 1.0000 - loss: 2.4814
OUTPUT >>> Embedding for 'great':
[ 0.019222   -0.01932465 -0.03971442 -0.02196687 -0.02987816
 0.00716395 -0.01269478 0.02227857 0.01485946 -0.07757436]
```

-0.01269478 0.02227857 0.01485946 -0.07757436]

To highlight it, the following is the desired ten-element word vector or embedding for the word "great" by using the CBOW model.

[0.019222   -0.01932465 -0.03971442 -0.02196687 -0.02987816
 0.00716395 -0.01269478 0.02227857 0.01485946 -0.07757436]

Skip-gram Model

Skip-gram models are also unsupervised methods. Like CBOW, they are also based on shallow neural networks with only one or two hidden layers. But skip-gram is exactly the opposite of CBOW. skip-gram models predict context words (surrounding words) given a target word. If we choose window size=2 in the sample sentence, "he is a great scholar", we get the following possible pairs.

- (he, is) (is, he)
- (is, a) (a, is)
- (a, great) (great, a)
- (great, scholar) (scholar, great)

Each pair has a central word as input and a context word as output. For the first pair (he, is), the central word is "he," and the context word is "is." When we decide which is the target, the remaining one automatically becomes the surrounding. Similarly, in (great, scholar), if we decide "great" as the central or target word, "scholar" automatically becomes the context or the output word. It is for a 2-gram model. In a 3-gram model, sequences of three consecutive words are formed.

If we choose a window size = 3, the valid 3-gram groups are as follows.

- (he, is, a), (he, a, is)
- (is, a, great), (is, great, a)
- (a, great, scholar), (a, scholar, great)

Each group consists of a central word with context before and after it. For instance, in the group (a, great, scholar), the central word is "great," and the context words are "a" and "scholar."

Next, let's create and run a sample neural network-based skip-gam model for demonstration (Listing 7-4), following these broad steps.

1. Import the libraries.
2. Define a corpus.
3. Initialize the tokenizer.
4. Generate a vocabulary.
5. Convert to sequences.
6. Create skip-gram pairs.

CHAPTER 7   ADVANCED PRAGMATIC TECHNIQUES AND SPECIALIZED TOPICS IN NLP

7. Prepare input/output data.

8. Build a model.

9. Compile a model.

10. Train model.

11. Extract the embeddings.

Listing 7-4. Code Demo of Skip-gram Model

```
#Running a shallow neural network program to demonstrate skip-gram model
# Import necessary libraries
# For numerical operations
import numpy as np
# For deep learning models
import tensorflow as tf
# For creating a linear model
from tensorflow.keras.models import Sequential
# Layers for model building
from tensorflow.keras.layers import Embedding, Dense, Flatten
# For tokenizing text
from tensorflow.keras.preprocessing.text import Tokenizer
# For generating Skip-gram pairs
from tensorflow.keras.preprocessing.sequence import skipgrams

# Sample corpus for demonstration
sentences = ["he is a great scholar", "a great scholar writes great papers"]

# Initialize tokenizer and fit on text
tokenizer = Tokenizer()
# Fit tokenizer on sample sentences
tokenizer.fit_on_texts(sentences)
```

CHAPTER 7   ADVANCED PRAGMATIC TECHNIQUES AND SPECIALIZED TOPICS IN NLP

```python
# Get vocabulary size
total_words = len(tokenizer.word_index) + 1

# Convert sentences to sequences of word indices
# Convert sentences to integer sequences
sequences = tokenizer.texts_to_sequences(sentences)

# Generate word pairs (center, context) for Skip-gram
# Initialize list to store Skip-gram pairs
skip_gram_pairs = []
for seq in sequences:
    # For each sentence sequence
    pairs, _ = skipgrams(seq, vocabulary_size=total_words,
    window_size=2)
    # Generate Skip-gram pairs
    skip_gram_pairs.extend(pairs)   # Add pairs to list

# Separate input (center) and output (context) words
input_words, context_words = zip(*skip_gram_pairs)
# Convert input words to array
input_words = np.array(input_words)
# Convert context words to array
context_words = np.array(context_words)

# Define Skip-gram model
model = Sequential()
# Embedding layer with vector size 10
model.add(Embedding(total_words, 10, input_length=1))
# Flatten the output
model.add(Flatten())
# Output layer for vocabulary-size classification
model.add(Dense(total_words, activation='softmax'))

# Compile model with categorical cross-entropy
# Set optimizer and loss function
```

```
model.compile(optimizer='adam', loss='sparse_categorical_
crossentropy')

# Train model on (input, context) pairs
model.fit(input_words, context_words, epochs=20, verbose=1)

# Get embeddings for a specific word, e.g., "great"
# Find index of word "great"
great_index = tokenizer.word_index['great']
# Get embedding vector for "great"
great_embedding = model.layers[0].get_weights()[0][great_index]
# Print embedding
print("Embedding for 'great':", great_embedding)
```
- Epoch 1/20

2/2 ──────────────── 1s 4ms/step - loss: 2.0738
Epoch 2/20
2/2 ──────────────── 0s 4ms/step - loss: 2.0708
....
Epoch 20/20
2/2 ──────────────── 0s 3ms/step - loss: 2.0389
OUTPUT >>> Embedding for 'great':
[0.01801779 -0.0028935 -0.04373218
0.06102643 -0.04138939 -0.05270927
   0.1154007   -0.08093932 -0.07753371 -0.02089559]

To highlight it, the following is the desired ten-element word vector or embedding for the word "great" by using the skip-gram model.

[0.01801779 -0.0028935 -0.04373218
0.06102643 -0.04138939 -0.05270927
0.1154007   -0.08093932 -0.07753371 -0.02089559]

You may like to compare this output to that of the CBOW model, calculated in the previous section.

CBOW and Skip-gram Code Demonstration Using Word2Vec

Let's run a Word2Vec-based program to demonstrate CBOW and skip-gram models. The following Word2Vec code implementation of these models is based on a pre-built, optimized algorithm for training word embeddings. In comparison, neural network-based programs have been manually created and trained to use customized shallow neural networks. Word2Vec is highly efficient and handles large corpora more effectively than custom-built models.

Our program Listing 7-5 follows the following broad steps.

1. Import libraries and prepare data.
2. Tokenize sentences and prepare a corpus.
3. Initialize the Word2Vec model for CBOW or skip-gram.
4. Train the Word2Vec model.
5. Access word embeddings.
6. Evaluate and use the embeddings.

Listing 7-5. Code to together demonstrate CBOW and skip-gram models

```
#Running a Word2Vec based program to together demonstrate CBOW
and skip-gram models
# Import necessary libraries
import gensim
from gensim.models import Word2Vec
from nltk.tokenize import word_tokenize
```

CHAPTER 7   ADVANCED PRAGMATIC TECHNIQUES AND SPECIALIZED TOPICS IN NLP

```python
# Sample corpus (list of sentences)
sentences = ["he is a great scholar", "a great scholar writes great papers"]

# Tokenize the sentences into words
tokenized_sentences = [word_tokenize(sentence.lower()) for sentence in sentences]

# CBOW Model (use skip_gram=False for CBOW)
cbow_model = Word2Vec(tokenized_sentences, vector_size=10, window=2, min_count=1, sg=0)
# sg=0 denotes CBOW model

# Train CBOW model
cbow_model.save("cbow_model.bin")

# Predict word embedding for 'great' using CBOW model
cbow_embedding = cbow_model.wv['great']
print("CBOW Embedding for 'great':", cbow_embedding)

# Skip-gram Model (use skip_gram=True for Skip-gram)
skipgram_model = Word2Vec(tokenized_sentences, vector_size=10, window=2, min_count=1, sg=1)
# sg=1 denotes Skip-gram model

# Train Skip-gram model
skipgram_model.save("skipgram_model.bin")

# Predict word embedding for 'great' using Skip-gram model
skipgram_embedding = skipgram_model.wv['great']
print("Skip-gram Embedding for 'great':", skipgram_embedding)
OUTPUT 1>>> CBOW Embedding for 'great':
[-0.00536227  0.00236431  0.0510335
 0.09009273 -0.0930295  -0.07116809  0.06458873
 0.08972988 -0.05015428 -0.03763372]
```

CHAPTER 7   ADVANCED PRAGMATIC TECHNIQUES AND SPECIALIZED TOPICS IN NLP

```
OUTPUT 2>>> Skip-gram Embedding for 'great':
[-0.00536227   0.00236431    0.0510335
 0.09009273  -0.0930295    -0.07116809  0.06458873
 0.08972988  -0.05015428   -0.03763372]
```

Compare the embeddings for 'great' based on CBOW and skip-gram models using the Gensim library, which provides an efficient code implementation of Word2Vec for training and using word embeddings.

GloVe: Global Vectors for Word Representation

GloVe is an unsupervised machine learning algorithm that was first presented in 2014 at a conference organized in Doha, Qatar. GloVe is used to generate vector representations (word embeddings) of words. GloVe contains pre-defined dense vectors for approximately over 6 billion words of English literature, including some general-use characters like commas, braces, and semicolons. Users can utilize a pre-trained GloVe embedding in various dimensions, such as 50-d, 100-d, 200-d, or 300-d vectors, depending on the availability of computational resources and the task requirements. Here, *d* represents *dimension*. 50-d would indicate a vector of size 50, and so on.

The main difference between Word2Vec and GloVe is that Word2Vec learns word embeddings (or word vectors) by predicting context words using skip-gram or CBOW models (utilizing large corpora). Its focus is on the local context. GloVe utilizes global word co-occurrence statistics to seize overall word relationships in its embeddings. GloVe model emphasizes global as well as local context. Word2Vec is generally faster and more efficient for context-specific NLP tasks because it dynamically captures word relationships during training. GloVe works great for capturing overall word relationships across an entire text dataset because it learns from global word co-occurrences (meaning how often words appear together across many sentences). This way, GloVe captures a broader understanding of word meanings and their relationships.

Like Word2Vec, the word embeddings provided by GloVe can be used in multiple NLP tasks like NER, machine translation, question-answering systems, document similarity, and document clustering.

Below is a code demo (Listing 7-6) that utilizes pre-trained GloVe embeddings to find similar words.

```
# Using pre-trained GloVe embeddings to find similar words.
# Necessary steps before you run the the code
```

1. Go to the GloVe project page.
2. Download the glove.6B.zip file. It contains word embeddings of multiple dimensions like 50d, 100d, and so on.
3. Unzip the file to see glove.6B.50d.txt.
4. Place glove.6B.50d.txt in the same directory as your code file, or specify the correct path in the code if it's located elsewhere.

Listing 7-6. Program Using GloVe embeddings to find similar words

```
# Import necessary libraries
import numpy as np  # For numerical operations
from sklearn.metrics.pairwise import cosine_similarity
# For similarity calculation

# Load GloVe embeddings
# Path to GloVe file (ensure this file is in the directory or
provide the correct path)
embeddings_index = {}  # Dictionary to store word vectors

# Open and read the GloVe file
with open('glove.6B.50d.txt', encoding='utf-8') as f:
    for line in f:
```

```python
        values = line.split()  # Split each line into components
        word = values[0]  # First item is the word
        vector = np.asarray(values[1:], dtype='float32')
        # Remaining items are vector values
        embeddings_index[word] = vector  # Add word and its vector to dictionary

# Check if "scholar" exists in the GloVe vocabulary
if "scholar" in embeddings_index:
    scholar_vector = embeddings_index["scholar"]  # Get the vector for "scholar"

# Calculate cosine similarity between "scholar" and all other words
similar_words = {}  # Dictionary to store words and their similarity scores
for word, vector in embeddings_index.items():
    similarity = cosine_similarity([scholar_vector], [vector])[0][0]  # Compute similarity
    similar_words[word] = similarity  # Add word and similarity score to dictionary

# Sort words by similarity score in descending order
sorted_similar_words = sorted(similar_words.items(), key=lambda x: x[1], reverse=True)

# Print top 10 words similar to "scholar"
print("Top words similar to 'scholar':")
# Skip the first word as it will be "scholar" itself
for word, similarity in sorted_similar_words[1:11]:
    print(f"{word}: {similarity}")
OUTPUT >>> Top words similar to 'scholar':
```

historian: 0.8646411895751953
philosopher: 0.8056567311286926
poet: 0.7976035475730896
author: 0.7962594032287598
professor: 0.7924675345420837
eminent: 0.7883749008178711
literature: 0.7770242094993591
theologian: 0.7625619173049927
sociologist: 0.7605342864990234
linguist: 0.7595699429512024

Note that similarity stands for how close in meaning two words are based on the cosine similarity between their vectors. Values close to 1 means words are closer in meaning.

Advances like BERT and GPT

There have been significant advances in NLP in recent years. BERT stands for Bidirectional Encoder Representations from Transformers. It was first introduced in 2018. The most popular model in the GPT category is ChatGPT. It was first released in 2022. Both BERT and GPT come under the category of large language models (LLM) and are based on transformer models. LLMs are trained on massive amounts of data and can perform various tasks, such as text translation and text generation, with minimal fine-tuning. This chapter focuses on BERT; an upcoming is dedicated to GPT-based models.

BERT is a kind of neural network based on transformer architecture. It's a pre-trained model that can be fine-tuned for specific NLP tasks by simply adding a new layer or a small network on top of BERT's architecture. The BERT model is available in various sizes to accommodate different computational resource capacities. The following are the BERT models used in practice.

CHAPTER 7   ADVANCED PRAGMATIC TECHNIQUES AND SPECIALIZED TOPICS IN NLP

- **BERT-Base** is a smaller BERT model. It consists of 12 layers and 110 million parameters. It strikes a balance between performance and speed for general NLP tasks.

- **BERT-Large** is a larger version of BERT. It consists of 24 layers and 340 million parameters. This BERT model offers higher accuracy, but it requires more computational resources.

- **DistilBERT** is a smaller and faster variant of BERT with around 66% of BERT's size. This BERT model maintains high accuracy while enabling faster processing time.

- **TinyBERT** is a compact BERT model. It is designed for fast performance and low memory usage. It has fewer network layers and hidden units.

- **ALBERT** is a lightweight version of BERT. It shares parameters across layers, thereby reducing model size while maintaining performance.

The BERT-Base model is a widely used general-purpose BERT model for a variety of NLP tasks. Its wide range of uses is justified because it maintains a reasonable performance and speed while consuming fewer computational resources.

The Hugging Face framework streamlines working with pre-trained models, such as BERT, GPT, and others. Listing 7-7 is a code demo on how the Hugging Face framework can be utilized for NLP tasks. For this purpose, we utilize the bert-base-uncased model. It is a version of BERT-Base, which does not differentiate between uppercase and lowercase letters. This book concentrates more on the usage of BERT and GPT rather than their internal architecture.

CHAPTER 7　ADVANCED PRAGMATIC TECHNIQUES AND SPECIALIZED TOPICS IN NLP

Listing 7-7. Code demo for sentiment analysis

```
"""

Code demo for sentiment analysis using the sentence "Ramesh
is my friend and he is a great scholar" with Hugging Face
and BERT.

"""
import torch
from transformers import BertTokenizer,
BertForSequenceClassification, pipeline

# Set random seed for reproducibility
torch.manual_seed(42)

# Load pre-trained BERT model and tokenizer
model = BertForSequenceClassification.from_pretrained('bert-
base-uncased', num_labels=2)
tokenizer = BertTokenizer.from_pretrained('bert-base-uncased')

# Create a sentiment-analysis pipeline
nlp = pipeline('sentiment-analysis', model=model,
tokenizer=tokenizer)

# Test sentence
sentence = "Ramesh is my friend and he is a great scholar."

# Get sentiment prediction
result = nlp(sentence)

# Print result
print(result)
OUTPUT: [{'label': 'LABEL_1', 'score': 0.5709643363952637}]
The model is predicting LABEL_1, meaning a positive sentiment,
using 57.09% confidence.
```

CHAPTER 7   ADVANCED PRAGMATIC TECHNIQUES AND SPECIALIZED TOPICS IN NLP

7.4.3 Case Studies and Applications

Let's examine a business case and demonstrate its code implementation.

The owner of a large customer service platform (a large e-commerce company) wants to analyze incoming customer queries with an aim to improve response times and accuracy. He is a technology enthusiast, and he wants to test BERT and Hugging Face in this customer service application. With the help of this application, he wants to categorize customer queries into returns, product inquiries, and complaints. It allows him to route specific customer queries directly to the specialized teams. He wants to use BERT to pick up the context and content of each query and provide targeted responses. He hopes with such an implementation, his establishment can streamline customer service, cut operational costs, and enable more efficient handling of high query volumes.

Our code demo (Listing 7-8) does the following tasks.

- Loads pre-trained BERT models and tokenizers
- Tokenizes sample data queries
- Sets up minimal training to fine-tune BERT on sample data
- Uses fine-tuned model in Hugging Face pipeline for predictions
- Maps prediction results to defined categories

Listing 7-8. Program for the business case

```
# Code demo for the business case.
# Install these libraries if not dopne already.
!pip install transformers torch
from transformers import BertTokenizer, BertForSequenceClassification, Trainer, TrainingArguments, pipeline
```

CHAPTER 7 ADVANCED PRAGMATIC TECHNIQUES AND SPECIALIZED TOPICS IN NLP

```python
import torch
from torch.utils.data import DataLoader, Dataset
import numpy as np

# Define sample data
data = [
    {"text": "I want to return my order", "label": 0},
    # returns
    {"text": "Can I know the delivery status?", "label": 1},
    # product inquiries
    {"text": "My product is defective", "label": 2}
    # complaints
]

# Map label indices to categories
label_map = {0: "returns", 1: "product inquiries", 2: "complaints"}

# Load pre-trained BERT tokenizer and model
tokenizer = BertTokenizer.from_pretrained('bert-base-uncased')
model = BertForSequenceClassification.from_pretrained('bert-base-uncased', num_labels=3)

# Tokenize and preprocess data
class SampleDataset(Dataset):
    def __init__(self, data, tokenizer):
        self.data = data
        self.tokenizer = tokenizer

    def __len__(self):
        return len(self.data)

    def __getitem__(self, idx):
        item = self.data[idx]
```

```python
        inputs = self.tokenizer(item["text"], truncation=True,
        padding='max_length', max_length=32, return_
        tensors="pt")
        inputs = {k: v.squeeze() for k, v in inputs.items()}
        # Squeeze to remove extra dimension
        inputs['labels'] = torch.tensor(item["label"],
        dtype=torch.long)
        return inputs

# Create Dataset and DataLoader
dataset = SampleDataset(data, tokenizer)

# Set up minimal training for fine-tuning
training_args = TrainingArguments(
    output_dir="./results",
    evaluation_strategy="no",
    per_device_train_batch_size=2,
    num_train_epochs=20,
)

trainer = Trainer(
    model=model,
    args=training_args,
    train_dataset=dataset,
)

# Train the model (this is minimal; actual fine-tuning requires more data and epochs)
trainer.train()

# Use Hugging Face pipeline for prediction
nlp_pipeline = pipeline("text-classification", model=model,
tokenizer=tokenizer)
```

CHAPTER 7   ADVANCED PRAGMATIC TECHNIQUES AND SPECIALIZED TOPICS IN NLP

```
# Predict categories for new queries
new_queries = ["I received the wrong item", "Where is my package?", "Can I return this?", "I received a defective product"]
for query in new_queries:
    result = nlp_pipeline(query)[0]
    predicted_label = int(result['label'].split('_')[-1])
    # Get label index from 'LABEL_0', 'LABEL_1', etc.
    category = label_map[predicted_label]   # Map label to category
    print(f"Query: '{query}'\nPredicted Category: {category}\n")
```

Query: 'I received the wrong item'
Predicted Category: complaints

Query: 'Where is my package?'
Predicted Category: product inquiries

Query: 'Can I return this?'
Predicted Category: product inquiries

Query: 'I received a defective product'
Predicted Category: complaints

In this code, Hugging Face is used to do the following.

- Loads the pre-trained BERT model
- Provides a tokenizer for text preprocessing
- Sets up text classification (pipeline function)
- Encapsulates the prediction process in a single call
- Maps results to the defined categories

334

CHAPTER 7   ADVANCED PRAGMATIC TECHNIQUES AND SPECIALIZED TOPICS IN NLP

7.5 Chapter Recap

This chapter explored the essence of pragmatic analysis. We mainly focused on discourse integration, distributional semantics, and word embeddings. We discussed these broad topics in depth from conceptual, implementation, and practical applications points of view. These insights can help you apply these practices and your future NLP endeavors.

The current NLP focus areas are understanding multimodality, platform-specific communication styles, identity and community building, and politeness dynamics in digital interactions. Multimodality concentrates on how text, images, videos, and emojis work together to communicate meaning online. Popular web platforms like X and Instagram have developed their own style of communicating, using language, and sharing ideas. Some leading researchers are focusing on how each platform influences users' communication styles. The current NLP research areas also focus on politeness and ethical aspects of natural languages. All this research is focused on online and social media communications between groups of people. Current trends suggest that future NLP researchers will likely focus more on areas such as AI in communication, cross-cultural interaction, the ethics of interaction, technological advancements, and digital literacy.

7.6 Reference

[1]   Sharma, P., and Nagashree, N. (2022). "Survey on Natural Language Processing and its Applications." *International Journal of Computational Learning & Intelligence* (2), 12–14.

CHAPTER 8

Transformers, Generative AI, and LangChain

8.1 Introduction to Transformers in NLP

The transformer architecture was introduced in 2017 by Vaswani et al. in their groundbreaking paper, "Attention Is All You Need."

This paper revolutionized the fields of neural network architectures and natural language processing (NLP). This introductory paper on transformer architectures explores the basic architecture and workings of transformers. Transformers use self-attention mechanisms to change one complete sentence into another. Transformer models are fundamentally different from conventional deep learning models, such as RNNs and LSTMs, which process a text sentence word by word. Take the example of a simple English sentence, "I left my book on the left side of the table." The word *left* has entirely different meanings in the same sentence. As RNNs and LSTMs process the sentence word by word, they may fail to pick up the context-specific meanings of the word *left*. Transformers process entire sentences using self-attention, which enables them to understand any input text more quickly and accurately grasp the context-specific meaning of words. This way, transformer models can efficiently handle comparatively long sentences and complex relationships in any given input text. By the way, self-attention helps models to work out the important words (for meaning), no matter how far they are apart in the input sentence. Self-attention mechanisms help the models to better handle NLP tasks that involve language translation and text generation.

Let's briefly discuss the basics of the otherwise complex architecture of a transformer model. It comprises mainly an encoder, decoder, self-attention mechanism, and feed-forward neural network layers. If you need the full details, we suggest you refer to the original paper that introduces the transformer models to the world (see Figure 8-1).

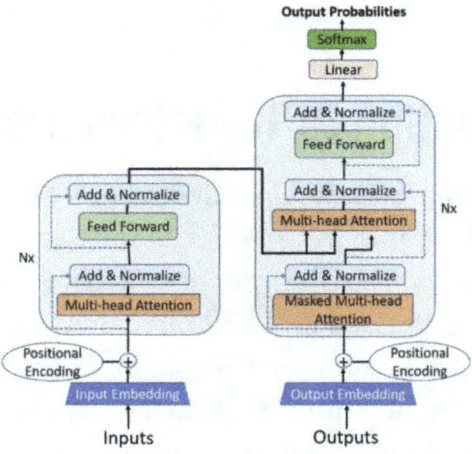

Figure 8-1. *Transformer model architecture (source)*

Before discussing the four layers of a transformer model, let's focus on the following basic terminology in this connection.

- **Self-attention**: It helps the model to have a global look at all the words in any input sentence to understand how the words in the sentence relate to each other in

CHAPTER 8   TRANSFORMERS, GENERATIVE AI, AND LANGCHAIN

the proper context. It helps the transformer model to focus on important words, irrespective of their position in the input sentence.

- **Cross-attention**: It facilitates the transformer model to connect the input sentence to the output sentence being generated by the model. Cross-attentions ensure the output sentence is created based on the right parts of the input sentence.

The following is a brief description of the basic elements of a transformer model.

- **Embedding**: It converts the given input words into numerical vectors (also called embeddings) that can be interpreted (understood) by the transformer model.

- **Encoder**: The input is put in the form of sentences to the encoder. It processes it through layers. Each of these layers employs a self-attention mechanism to focus on important words in any given input sentence. Then, the same self-attention mechanism takes the input sentence through the feed-forward networks. This whole process transforms the data into meaningful representations.

- **Decoder**: The final output sentence is generated by the decoder. It takes the input from the processed data of the encoder. The decoder also employs the self-attention mechanism to comprehend the generated words. It uses cross-attention to line up with the encoder's input.

- **Self-attention mechanism:** (This concept was discussed before introducing the model layers.)

- **Feed-forward neural network**: It is a basic artificial neural network (ANN) that works with each encoder and decoder layer to transform input data. This helps identify complex patterns in the data as it is processed.

To summarize the workings of transformer models, all these elements work together to convert any given input sentence into an output sentence. It is the job of the embedding layer to convert input words to numerical vector formats. Then comes the encoder that uses self-attention to retrieve the important words, their relationships, and the context. Feed-forward networks help encode in this process that finally extracts meaningful patterns from the input sentence. Next comes the decoder, which takes inputs from the data processed by the encoder to refine the output sentence step by step. Elements such as cross-attention and self-attention aid the decoder in this process. Now comes the output layer that transforms the processed data into the final words that are presented in the form of the output sentence.

8.1.1 Evolution of NLP: From Traditional Models to Transformers

The first significant foundational stone in NLP was laid by the Georgetown-IBM experiment in 1954, which successfully translated 60 Russian sentences into English using rule-based techniques. Until the 1970s, it was considered an era of rule-based NLP systems, mainly deployed for machine translation. During this period, researchers developed complex sets of linguistic rules, enabling computers of the era to process human language. The 1980s and 1990s are marked as an era of development in statistical methods and machine learning algorithms for the machine processing of human languages. Hidden Markov Models and support vector machines especially prevailed during this period in the NLP domain. These techniques allowed more flexible and data-driven approaches to NLP.

CHAPTER 8 TRANSFORMERS, GENERATIVE AI, AND LANGCHAIN

From the early 2000s to the present, deep learning-based models have become mainstream for processing human languages. In fact, neural networks have revolutionized the field of NLP by enabling complex tasks such as speech processing, improved machine translations, and social media sentiment analysis. Neural network models, such as recurrent neural networks (RNNs), long short-term memory (LSTM) networks, and transformers have played a key role in these advancements. Figure 8-2 depicts these advances in the NLP domain from the 1950s until 2020. From 2020 onward, we have all witnessed the emergence of game-changing GenAI models, such as ChatGPT, and many more.

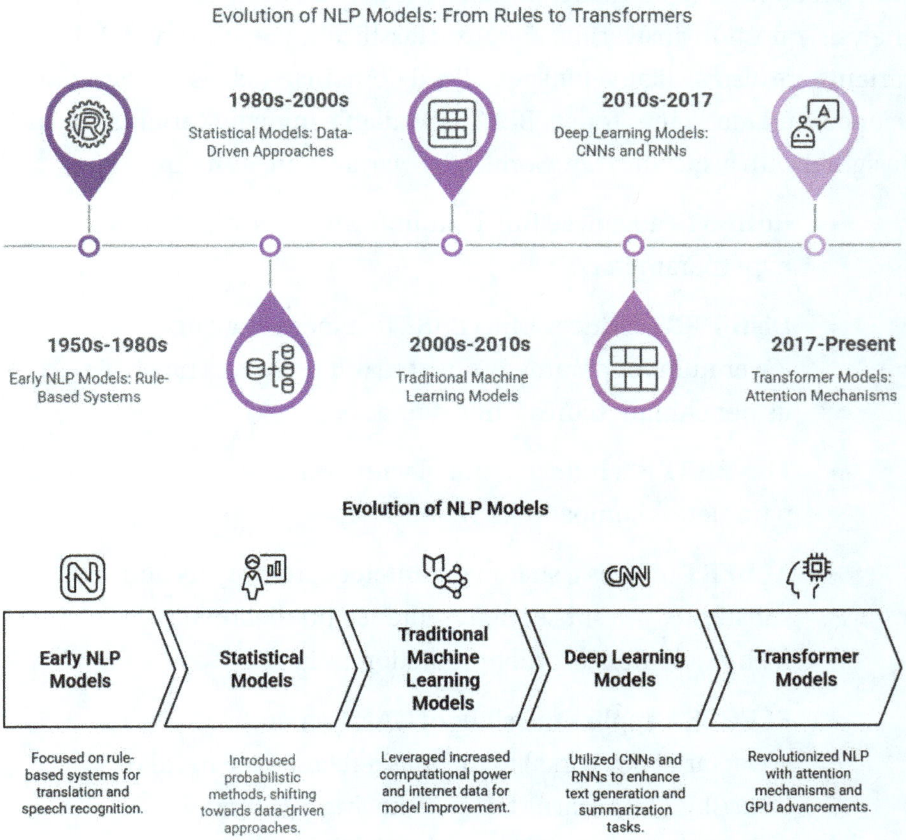

Figure 8-2. The evolution of NLP models from translation (early 1950s) to transformers (2017)

8.1.2 Overview of BERT by Google

BERT (Bidirectional Encoder Representations from Transformers) was made available for general public usage in 2018. BERT was released as an open-source license by Google. BERT is a widely used transformer-based model. It is grounded on the transformer model architecture, explicitly its encoder module. BERT is designed to recognize the context of words in sentences. BERT processes words in the context of entire input sentences and does not look at words one by one, as done by some techniques employed for NLP tasks. This way, BERT can pick up the word's meaning even if it appears to be vague. BERT excels in tasks such as sentiment analysis, question answering, and text classification. Fine-tuned BERT variants are also available for specialized domains, such as biology, data science, and medicine. Today, BERT is available in many variants, each designed with a specific trait. Some of its variants are as follows.

- **RoBERTa** advances BERT's training by altering vital hyperparameters.

- **DistilBERT** reduces BERTBASE to a model having fewer number parameters, whereas it preserves most of its benchmark scores while doing so.

- **TinyBERT** is a reduced model with even fewer parameters compared to DistilBERT.

- **ALBERT** utilizes a shared parameter across layers and substitutes the subsequent sentence prediction task with the sentence-order prediction task.

- **ELECTRA** applies the clues of GAN (generative adversarial networks) to the masked language model task. ELECTRA learns faster by having one model create fake words and another model detect if they are real or fake.

RoBERTa is widely considered suitable for general-purpose NLP tasks. It is the best for tasks that require high accuracy and strong language comprehension. To use RoBERTa, resource availability should not be a constraint. DistilBERT is much smaller in size than RoBERTa, and it is the most popular variant of BERT because it balances finely between performance and efficiency. DistilBERT needs much fewer computational resources (than RoBERTa) and still maintains around 97% of BERT's accuracy. Due to these advantages, DistilBERT is used for many practical applications across domains. DistilBERT is especially useful when resources are limited. Our code demo uses DistilBERT as the sample program is being developed and run on a modest business computer.

Note Masked language modeling (MLM) is a technique where certain words in an input sentence are randomly masked. The model is then trained to predict the missing words based on the surrounding context. A GAN is an AI variant that uses two competing models: one produces false data and the other detects if the fake data produced by the first model is real or fake. ELECTRA advances learning speed by substituting masked words with fake ones and then training the model to spot them. It permits ELECTRA to learn from every word and not by just the limited words it guesses.

8.1.3 Code to Demonstrate the Use of DistilBERT Using PyTorch

Listing 8-1 depicts the Code Example on How to Use DistilBERT with PyTorch

Listing 8-1. Making use of Use of DistilBERT with PyTorch

```
# Install the following libraries if not done already.
#!pip install transformers datasets torch
# Install PyTorch for CPU-only support
```

CHAPTER 8   TRANSFORMERS, GENERATIVE AI, AND LANGCHAIN

```python
#!pip install torch torchvision torchaudio!
#!pip install --user "accelerate>=0.26.0"
# For GPU Support (NVIDIA CUDA)
#!pip install torch torchvision torchaudio --index-url https://
download.pytorch.org/whl/cu118
#!pip install --user "accelerate>=0.26.0"
# This program shows how to fine-tune and use the DistilBERT
model for sentiment analysis on the IMDb dataset..
# It involves:
# 1. Loading the IMDb dataset and preparing it.
# 2. Tokenizing the text data using DistilBERT's tokenizer.
# 3. Fine-tuning the pre-trained DistilBERT model for binary
classification.
# 4. Evaluating the fine-tuned model and making predictions on
new text.

# Explanation of imports:
# - DistilBertTokenizer: Used to tokenize input text into the
format required by the DistilBERT model.
# - DistilBertForSequenceClassification: Pre-trained DistilBERT
model tailored for sequence classification tasks.
# - Trainer: High-level API for training and evaluating Hugging
Face models.
# - TrainingArguments: Configuration for the Trainer, including
batch size, epochs, learning rate, etc.
# - load_dataset: Part of the `datasets` library, used to load
and manage datasets like IMDb for training and evaluation.
from transformers import DistilBertTokenizer,
DistilBertForSequenceClassification, Trainer, TrainingArguments
from datasets import load_dataset
# Final code follows
# Step 1: Load Dataset
```

```python
from datasets import load_dataset
dataset = load_dataset("imdb")

# Split into training and validation sets
train_data = dataset["train"].shuffle(seed=42).
select(range(2000))   # Select smaller subset for demo
val_data = dataset["test"].shuffle(seed=42).select(range(500))
# Step 2: Load Tokenizer
from transformers import DistilBertTokenizer
tokenizer = DistilBertTokenizer.from_pretrained("distilbert-base-uncased")

# Tokenize the data
def tokenize_function(examples):
    return tokenizer(examples["text"], padding="max_length",
    truncation=True)

train_data = train_data.map(tokenize_function, batched=True)
val_data = val_data.map(tokenize_function, batched=True)

# Set format for PyTorch
train_data.set_format(type="torch", columns=["input_ids",
"attention_mask", "label"])
val_data.set_format(type="torch", columns=["input_ids",
"attention_mask", "label"])

# Step 3: Load Model
from transformers import DistilBertForSequenceClassification
model = DistilBertForSequenceClassification.from_pretrained(
    "distilbert-base-uncased", num_labels=2
)

# Step 4: Define Training Arguments
from transformers import TrainingArguments
```

```python
training_args = TrainingArguments(
    output_dir="./results",
    evaluation_strategy="epoch",  # 'eval_strategy' for updated versions
    learning_rate=2e-5,
    per_device_train_batch_size=16,
    per_device_eval_batch_size=16,
    num_train_epochs=3,
    weight_decay=0.01,
    logging_dir="./logs",
    logging_steps=10,
    save_strategy="no"
)

# Step 5: Trainer Object
from transformers import Trainer, TrainingArguments
trainer = Trainer(
    model=model,
    args=training_args,
    train_dataset=train_data,
    eval_dataset=val_data,
    processing_class=tokenizer,  # Updated to 'processing_class' to avoid deprecation warning
)

# Step 6: Train and Evaluate
trainer.train()
trainer.evaluate()

# Step 7: Inference
def predict(text):
    tokens = tokenizer(text, return_tensors="pt",
    truncation=True, padding="max_length", max_length=512)
```

CHAPTER 8　TRANSFORMERS, GENERATIVE AI, AND LANGCHAIN

```
    outputs = model(**tokens)
    prediction = outputs.logits.argmax(-1).item()
    label = "Positive" if prediction == 1 else "Negative"
    return label

# Example prediction 1
text = "The movie was fantastic! The plot and acting were
top-notch."
print(f"Review: '{text}'\nPrediction: {predict(text)}")
```
Some weights of DistilBertForSequenceClassification were not initialized from the model checkpoint at distilbert-base-uncased and are newly initialized: ['classifier.bias', 'classifier.weight', 'pre_classifier.bias', 'pre_classifier.weight']
You should probably TRAIN this model on a down-stream task to be able to use it for predictions and inference.
<IPython.core.display.HTML object>
<IPython.core.display.HTML object>
Review: 'The movie was fantastic! The plot and acting were top-notch.'
Prediction: Positive
```
# Example prediction 2
text = "The movie was super boring! The plot and acting were
terrible."
print(f"Review: '{text}'\nPrediction: {predict(text)}")
```
Review: 'The movie was super boring! The plot and acting were terrible.'
Prediction: Negative
```
# Example prediction 3
text = "This morning I am not able to decide what to do with
the day."
print(f"Review: '{text}'\nPrediction: {predict(text)}")
```

Review: 'This morning I am not able to decide what to do with the day.'
Prediction: Negative
# Example prediction 4
text = "The movie is neighther good nor bad."
print(f"Review: '{text}'\nPrediction: {predict(text)}")
Review: 'The movie is neighther good nor bad.'
Prediction: Negative
# Example prediction 4
text = "The movie was neutral."
print(f"Review: '{text}'\nPrediction: {predict(text)}")
Review: 'The movie was neutral.'
Prediction: Negative
# Step 8: Save the Model and Tokenizer
output_dir = "C://AA SK 53//A INDUS//Papers with Murugan///NLP Book//Python Tutorials//Chapter 8//saved_model_8.1"
model.save_pretrained(output_dir)
tokenizer.save_pretrained(output_dir)

print(f"Model and tokenizer saved to {output_dir}")
Model and tokenizer saved to C://AA SK 53//A INDUS//Papers with Murugan///NLP Book//Python Tutorials//Chapter 8//saved_model_8.1
Load the saved model and make predictions
from transformers **import** DistilBertForSequenceClassification, DistilBertTokenizer

# Step 1: Load the saved model and tokenizer
output_dir = "C://AA SK 53//A INDUS//Papers with Murugan///NLP Book//Python Tutorials//Chapter 8//saved_model_8.1//"
loaded_model = DistilBertForSequenceClassification.from_pretrained(output_dir)

```
loaded_tokenizer = DistilBertTokenizer.from_
pretrained(output_dir)
```

```
# Step 2: Define the prediction function
def predict_with_loaded_model(text):
    tokens = loaded_tokenizer(text, return_tensors="pt",
    truncation=True, padding="max_length", max_length=512)
    outputs = loaded_model(**tokens)
    prediction = outputs.logits.argmax(-1).item()
    label = "Positive" if prediction == 1 else "Negative"
    return label
```

```
# Step 3: Make a prediction
example_text = "The movie was boring and lacked excitement."
print(f"Review: '{example_text}'\nPrediction: {predict_with_
loaded_model(example_text)}")
Review: 'The movie was boring and lacked excitement.'
Prediction: Negative
```

BERT is based on the concept of transfer learning, which is like first learning a bicycle and then transferring the learnings on the bicycle to learn a bike. Naturally, if you know how to ride a bicycle, learning a motorbike or a scooter becomes a lot easier. In the technology language, transfer learning is as follows. You first train a large NLP model on very large text corpora (large text collections), including Wikipedia and many more. This was the large NLP model under consideration here, which learns how words and sentences (in a huge training text) connect. Once the NLP model develops the understanding of language (like English) in general, later we can slightly tweak the already trained NLP model to perform specific tasks like finding a text's emotion (popularly known as sentiment analysis) or making a question answering system in the domains like finance, tourism, healthcare and many more.

BERT is a great example of a system working on the concepts of transfer learning. BERT is trained on enormous text data. This way, it develops a great understanding of the language in which it is trained. Later, we can slightly alter BERT and train it with our data (new problem-specific data) to efficiently perform NLP tasks like text classification, named entity recognition, language translation, summarization, and paraphrase detection. This retraining or tweaking of a trained BERT model can be done using a very small amount of data.

8.1.4 Multimodal NLP: Combining Text with Images, Audio, and More

Multimodal NLP combines diverse forms of information, such as text, speech, images, and videos, as inputs to make NLP tasks smarter. By combining different types of information as input, machines can better understand the communication between humans, as humans are also having similar capabilities. For example, to comprehend the class lessons, one may utilize class lectures, written texts, pictures, and videos for support. Another example of this is in face-to-face meetings. Facial expressions are also available, which can help better in another person's mood detection. In a similar fashion, machines can enhance language tasks, such as translation, by incorporating pictures and videos to convey meanings more effectively. Multimodal NLP is a relatively new area of research.

The following are some of the tools available for multimodal NLP.

- **OpenAI CLIP** links images and text. Its usage is mainly in image search or visual language understanding.

- **Google Vision transformer (ViT)** links vision and language tasks. This tool refines the model's performance in image classification and the generation of automatic captions.

- **Multimodal BERT (ViLBERT)** extends BERT so that it can process both visual and textual data. This tool is especially useful for NLP tasks such as visual question answering and automatic image captioning.

- **DeepMind Perceiver** is a versatile model capable of handling various input data types, including text, images, and audio. It is highly effective in multimodal learning.

- **TensorFlow Multimodal Toolkit (TF-MMT)** is an all-inclusive framework for developing and training multimodal NLP models. It is useful for NLP tasks such as emotion detection and the analysis of multimedia content.

8.1.5 Demonstrating the Use of OpenAI CLIP for Multimodal Understanding

The theme of Listing 8-2 demonstrates the use of OpenAI CLIP for multimodal understanding, specifically to compute text-to-image similarity. The program assesses the degree to which various text descriptions align with a given image by utilizing CLIP's capability to integrate textual and visual representations.

Listing 8-2. Code to Demonstrate the use OpenAI CLIP for Multimodal Understanding

```
import torch
from transformers import CLIPProcessor, CLIPModel
from PIL import Image
import matplotlib.pyplot as plt
```

CHAPTER 8 TRANSFORMERS, GENERATIVE AI, AND LANGCHAIN

```python
# Load the CLIP model and processor
model = CLIPModel.from_pretrained("openai/clip-vit-base-patch16")
processor = CLIPProcessor.from_pretrained("openai/clip-vit-base-patch16")

# Function to display the image
def display_image(image_path):
    img = Image.open(image_path)
    plt.imshow(img)
    plt.axis('off')
    plt.show()

# Example image and text for image-to-text search
image_path = 'dog_with_person_on_beach.jpg'  # Replace with your image path
text_descriptions = ["A dog running on the beach",
                     "A golden retriever in a park",
                     "A person surfing on a beach",
                     "A sunset over the ocean"]

# Preprocess image and text
image = Image.open(image_path)
inputs = processor(text=text_descriptions, images=image, return_tensors="pt", padding=True)

# Get image and text features
with torch.no_grad():
    outputs = model(**inputs)

# Calculate similarity
image_features = outputs.image_embeds
text_features = outputs.text_embeds
similarities = torch.cosine_similarity(image_features, text_features)
```

CHAPTER 8 TRANSFORMERS, GENERATIVE AI, AND LANGCHAIN

```
# Show similarity scores
for i, text in enumerate(text_descriptions):
    print(f"Text: '{text}' | Similarity: {similarities[i]:.4f}")

# Display the image
display_image(image_path)
Text: 'A dog running on the beach' | Similarity: 0.3106
Text: 'A golden retriever in a park' | Similarity: 0.1987
Text: 'A person surfing on a beach' | Similarity: 0.2373
Text: 'A sunset over the ocean' | Similarity: 0.2173
```

Output of Listing 8-2, "A dog running on the beach," has the highest similarity score (0.3106), indicating it best matches the image. This suggests the image most likely depicts a scene closely related to the given description, highlighting CLIP's effectiveness in text-to-image relevance.

8.1.6 Text-to-Image Search Using OpenAI's CLIP Model

Listing 8-3 demonstrates text-to-image search using OpenAI's CLIP model. It matches a given text description with a set of images by computing cosine similarity between their embeddings. The most relevant image is then identified and displayed, highlighting CLIP's capability to understand the connection between text and images.

CHAPTER 8 TRANSFORMERS, GENERATIVE AI, AND LANGCHAIN

Listing 8-3. Program for Text-to-Image Search Using CLIP Model

```
import torch
from transformers import CLIPProcessor, CLIPModel
from PIL import Image

# Load the CLIP model and processor
model = CLIPModel.from_pretrained("openai/clip-vit-base-patch16")
processor = CLIPProcessor.from_pretrained("openai/clip-vit-base-patch16")

# Example images and text description for text-to-image search
images = ["dog_with_person_on_beach.jpg", "empire_state.jpg", "running_dog_on_beach.jpg", "dog_play.jpg", "example_image.jpg"]
text_description = "A dog running on the beach in the sunshine"

# Preprocess the text
inputs_text = processor(text=[text_description], return_tensors="pt", padding=True)

# List to store similarities for each image
image_similarities = []

# Loop through images
for image_path in images:
    # Open image
    image = Image.open(image_path)

    # Preprocess the image and get features
    inputs_image = processor(text=[text_description], images=image, return_tensors="pt", padding=True)
    with torch.no_grad():
        outputs = model(**inputs_image)
```

```
# Calculate similarity
image_features = outputs.image_embeds
text_features = outputs.text_embeds
similarity = torch.cosine_similarity(image_features, text_features)

image_similarities.append((image_path, similarity.item()))
```

```
# Sort images by similarity score
sorted_images = sorted(image_similarities, key=lambda x: x[1], reverse=True)
```

```
# Display the most relevant image
best_image_path = sorted_images[0][0]
# Assuming display_image is a function to display the image.
Replace with your own method if needed.
display_image(best_image_path)
```

```
print(f"The most relevant image based on the description is: {best_image_path}")
```

The most relevant image based on the description is running_dog_on_beach.jpg. Output of Listing 8-3.

8.2 Overview of GPT

GPT is the acronym for *generative pre-trained transformer*, a noteworthy leap in NLP and AI in general. The model works with advanced machine learning techniques and generate human-like conversations when appropriate prompts are given. GPT family of models like GPT-2, GPT-3, and GPT-4 have come up (comparatively) recently with the development efforts of OpenAI. These models demonstrate remarkable capabilities in NLP tasks, including text generation, summarization, and machine translation. These and many other such models (by multiple companies) are popularly known as GenAI (generative AI) models.

Let's start with a brief paragraph on the evolution of GenAI systems, starting with RNNs.

8.2.1 Evolution of Text Generation Models: From RNNs to GPT

RNNs (early 1980s) were one of the significant steps in text processing. RNNs are capable of processing text serially while preserving hidden steps across many timesteps. RNNs typically struggle with processing long queues of text data (long-term dependencies due to the problem of vanishing gradients). The development of LSTMs and GRUs, as improvements over RNNs, enables them to handle long-term dependencies. Then, Seq2Seq models with attention mechanisms were introduced, allowing NLP models to focus on specific important parts of the input text. This innovation revolutionized NLP tasks, including text summarization and machine translation. Despite these milestone developments, NLP models still lacked the capability to handle longer contexts, and they were not scalable to the required extent. This book does not delve into the fine details of neural networks and deep learning.

CHAPTER 8   TRANSFORMERS, GENERATIVE AI, AND LANGCHAIN

Following a breakthrough in the development of transformer models in 2017, text processing, particularly text generation, has made significant progress. The self-attention mechanisms of transformer models not only enabled parallel processing but also made it possible to deal with longer contexts, a capability that was missing in the earlier models. The introduction of transformer models laid a strong foundation for the development of GPT models by OpenAI. GPT models were pre-trained on massive NLP datasets, and they could be fine-tuned for specific tasks. A few years later, the development of GPT2, GPt-3, and more recently GPT-4 scaled up the capabilities of earlier such models by manifolds. These models are state-of-the-art when it comes to consistency and reasoning. Figure 8-3 depicts this evolution.

CHAPTER 8   TRANSFORMERS, GENERATIVE AI, AND LANGCHAIN

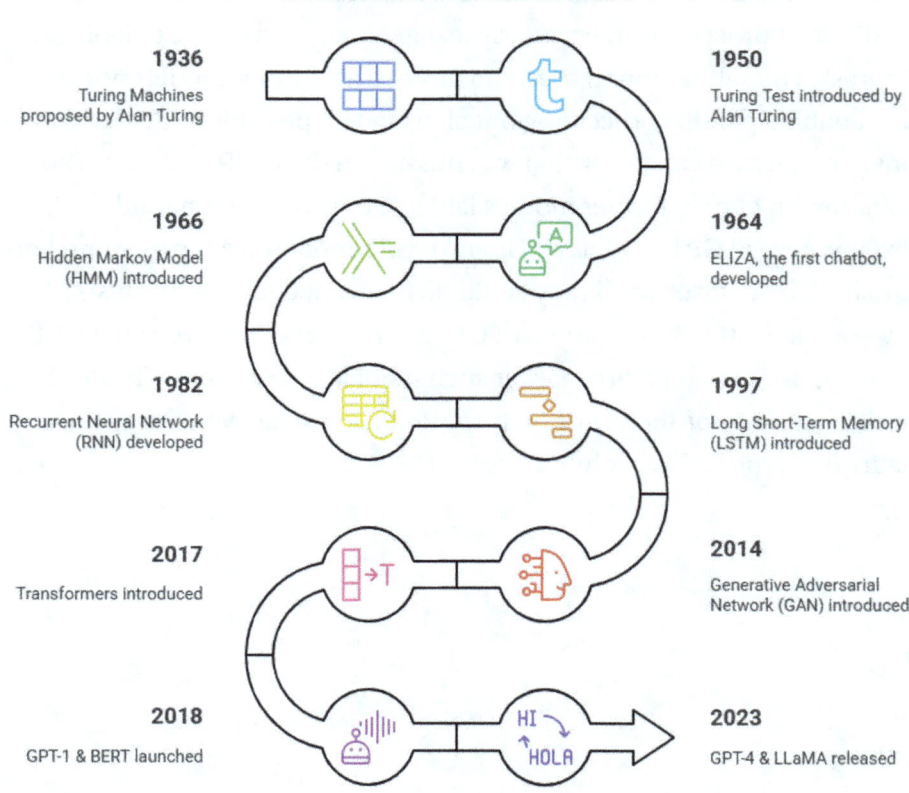

Figure 8-3. *The evolution of NLP models from RNNs to GPT-4*

More recently, many other AI-focused companies, like Meta (Facebook's parent company), have come up with their text generation models. Despite this, OpenAI's models remain one of the most popular.

8.2.2 A Short Note on the Working of GenAI (LLMs) Models

GenAI models, like LLMs, are based on cutting-edge neural networks trained to generate human-like text. GenAI models use a transformer architecture that depends on self-attention mechanisms, and they use embeddings to process the input text in a given context. These models analyze large datasets to learn patterns, grammar, and even factual knowledge. It enables them to accomplish NLP tasks like predicting the next word in a sentence or articulating (generating) text from prompts. Examples of GenAI models contain GPT and BERT-based LLMs.

These models first transform the input text into numerical vectors. These vectors are passed through multiple layers of the advanced neural network. Each layer works to refine the understanding of the text. These layers can capture relationships between words and phrases of the input text. The attention mechanisms in the model focus on the most relevant parts of the input. These mechanisms ensure a context-aware output. Training GenAI models involves predicting masked or next words, and further refining the fluency and accuracy of the models.

GenAI models contain billions of parameters (weights and biases). The enormous size of the parameters makes the models capable of diverse tasks, such as text summarization, translation, and creative writing. While GenAI models are extremely effective, they need vast computational resources. They can sometimes produce biased or factually incorrect outputs (called hallucinations) depending on the training data. A more detailed theoretical explanation of the workings of GenAI models is beyond the scope of this volume; the main focus of this book remains the application of concepts rather than going in-depth into the underlying theory.

8.2.3 Applications and Future Trends in Generative AI

The applications of GenAI models like ChatGPT are literally countless. These applications span diverse industries, including healthcare, hospitality, information technology, manufacturing, banking and finance (BFSI), advertising, marketing, sales, and media and entertainment.

CHAPTER 8   TRANSFORMERS, GENERATIVE AI, AND LANGCHAIN

In fact, separate volumes are written on applications of GenAI (and AI in general) in functional areas.

In the healthcare and pharmaceutical industries, GenAI finds applications in all functional areas. These applications include drug discovery, personalized treatments, predictive diagnosis, enhancing medical images, and simplifying tasks with patient notes and information. The applications of GenAI in the marketing, sales, and advertising industries include generating marketing text and images, creating personalized recommendations, crafting product descriptions, and enhancing search engine optimization. In the manufacturing domain, GenAI is utilized for code generation, programming language translation, code documentation, and automated testing. The BFSI domain utilizes GenAI for creating profitable investment strategies, portfolio management, communicating and educating clients and investors, drafting documentation, and monitoring regulatory compliance. This book does not provide a detailed description of these applications due to space constraints; refer to any standard text on the application of AI in functional areas.

AI and large language models, commonly referred to as LLMs, have immense potential to assist humanity in a wide range of tasks. But at the same time, GenAI models demand their responsible and ethical usage. Generative AI can produce content that is far removed from reality, commonly referred to as a hallucination. Hallucination raises ethical concerns about GenAI, including those related to truth, trust, and privacy. Apart from this, other challenges associated with GenAI models include algorithmic bias, data privacy, copyright infringement, misinformation, and job displacement. Many research groups across the globe are working on the challenges of GenAI. The current trends in this direction include responsible development, transparency, and fairness.

Generative AI is evolving at a rapid rate in its quest to create output and interactions that are personalized and human-like. Researchers are exploring GenAI interactions with edge computing and the Internet of Things (IoT). This can aid in real-time and localized content creation.

CHAPTER 8   TRANSFORMERS, GENERATIVE AI, AND LANGCHAIN

In future AI-based products, we can expect reduced bias and lesser hallucination. With advancements in GenAI models, we can expect augmented creativity and capabilities to automate both complex and repetitive tasks. These developments in the near future are likely to advance efficiencies and reduce costs across the industry domains. Collaborating with AI tools and robots will become an everyday part of everyone's lives.

Note   Edge computing processes data close to where it's created. It is like processing localized data on your phone or a nearby computing device, rather than sending it to a remote server for processing. Let's take an example of premium smartwatches that are capable of analyzing your heartbeat without the need for the Internet. Now let's talk in brief about IoT. It is a complex system of interconnected devices, such as smart bulbs, cars, or machines, that communicate with each other and exchange data (both ways) with the help of the Internet. IoT devices can be present anywhere, including homes, vehicles, factories, and many more. For example, a smart thermostat (for a home central air conditioner) utilizes the Internet to periodically receive weather updates (all automatically) and regulate your home's temperature within the comfortable limits of its inhabitants.

8.3 LangChain and OpenAI's (GenAI) APIs

LangChain, when combined with the state-of-the-art GenAI APIs from OpenAI, signifies an influential way to build intelligent and dynamic applications for various business scenarios and complex workflows. LangChain is an open-source project launched in October 2022 by Harrison Chase. It offers a comprehensive framework for building advanced workflows by integrating large language models (LLMs). LangChain can seamlessly connect with LLMs from OpenAI (and others) along with external data, tools, and logic. This combination is capable of producing state-of-the-art conversational applications that generate human-like responses. These applications are also useful in NLP tasks, such as content generation and decision support.

As discussed, OpenAI APIs act as tools for developers to utilize advanced AI models, like ChatGPT, in their applications and workflows (that can manage multistep processes). Such applications can comprehend input texts to create desired content, develop question-answer type bots for specific domains, perform text summarization, and even assist in translations. OpenAI APIs are user-friendly, and help developers add smart AI features to routine business applications. Applications (apps) developed by combining OpenAI APIs with LangChain support memory can remember previous conversations and process steps, helping the apps to better handle contexts. It creates a perfect foundation for creating applications such as chatbots, interactive systems, and decision-making tools. LangChain also supports human-like trendy AI agents (sometimes also referred to as agentic AI) that can take autonomous actions to complete a variety of tasks, as discussed later in this chapter.

LangChain is fundamentally a modular structure architecture. It permits the seamless integration of a variety of components that include prompts, memories, and chains. These are LangChain-specific terms, which is discussed later in this chapter. The modularity aspect of LangChain allows the creation of flexible applications while preserving clarity and simplicity. Additionally, LangChain supports a range of LLMs, including those from OpenAI, Azure, Anthropic, and Google Cloud. OpenAI's LLMs, being one of the most popular, are featured in this book.

You need not be an AI expert to work with OpenAI APIs. Basic knowledge of programming with languages like Python is sufficient to produce trendy and sophisticated applications with AI features using these APIs.

8.3.1 Setting Up the Environment

Before you use LangChain to work with OpenAI functions, a suitable environment for development purposes needs to be created. The following four simple steps serve as a guide for you to create a Python environment with the required libraries installed.

1. **Create a virtual environment for your project.**
   First, you need to navigate to the project folder on your Windows machine and create the virtual environment using the following environment.

   Activate the environment using the following command on the prompt.

   Windows: `myenv\Scripts\activate`

   macOS/Linux: `source langchain-env/bin/activate`

   We use the Anaconda Python distribution, so we executed these commands on the Anaconda Prompt.

2. **Install required packages.** Run the following command on the prompt to install LangChain and OpenAI Python client. All the following commands are run the same way.

3. **Get an OpenAI key. An** OpenAI key is required to seamlessly interact with the OpenAI models. Follow these steps.

   a. Open an OpenAI account on the OpenAI official website.

   b. Navigate to the API section on your dashboard, where you can generate and also manage your OpenAI API keys.

   c. Keep these API keys for the security reasons.

4. **Set the API key in the required environment.** To get the OpenAI API keys, follow these simple steps.

   a. Hardcoding the API key into your code is not advisable for the security and integrity of your newly created key.

   b. The best practice is to set your API key as an environment variable as follows.

   c. Execute the following commands on your terminal. It temporarily sets the environment variable for the existing terminal session.

   Windows: `set OPENAI_API_KEY='my-openai-api-key'`

   macOS/Linux: `export OPENAI_API_KEY='my-openai-api-key'`

Congratulations! Now, you are ready to create your favorite application using LangChain and OpenAI models. We develop the following sections step by step to create your dream AI applications.

8.4 Model I/O: Easily Interface with Language Models

Model input/output (I/O) serves as a framework within the LangChain system. It provides a structured approach to interacting with LLMs, such as OpenAI and others. Model I/O controls the flow of data to and from the models. It includes preparing input prompts and processing model outputs. This section focuses on the basic functions and syntax of LangChain to demonstrate how to create input prompt requests and handle the model outputs effectively. Advanced techniques like few-shot prompt templates are also covered. Few-shot prompt templates improve model responses by providing contextual examples. Additionally,

CHAPTER 8   TRANSFORMERS, GENERATIVE AI, AND LANGCHAIN

we discuss serialization methods for efficiently saving and loading prompts. By mastering these tools, you can build more robust and flexible applications that leverage the full potential of advanced AI models like ChatGPT from OpenAI.

Listing 8-4 is our first demo program portraying LangChain with OpenAI chat models. Chat models are more popular, so we are taking it as our first demo program.

```
""" In the following demo program, you may find many unknown
terms. We have tried to explain them
within the program itself. No worries but! We will take up all
of it in the below sections."""
```

Listing 8-4. Building a Dynamic Chat Assistant with LangChain and OpenAI's Chat Models

```
"""This program shows how to use LangChain and OpenAI's chat
models to create a smart assistant. It combines user input with
instructions for the AI to give helpful answers."""
# You could store the openai key in the form of a string in a
.txt file
# This way of storing the API key is not very secure, but
suitable for your personal projects on local computers
# Below example will make it more clear for you
# Install the essential libraries if not done already
# !pip install openai langchain
f = open('C:\\Users\\Shailendra Kadre\\Desktop\\OPEN_AI_
KEY.txt')
api_key = f.read()
# Test of api_key is copied correctly
# api_key
from langchain.chat_models import ChatOpenAI
```

CHAPTER 8 TRANSFORMERS, GENERATIVE AI, AND LANGCHAIN

```python
from langchain.prompts.chat import (
    ChatPromptTemplate,
    HumanMessagePromptTemplate,
    SystemMessagePromptTemplate
)
from langchain.schema import AIMessage, HumanMessage, SystemMessage
"""
Step 1: Import Necessary Modules
- `ChatOpenAI`: Enables interaction with OpenAI's language models like GPT-3.5 Turbo.
- `ChatPromptTemplate`: Combines system and human message templates into a conversation format.
- `HumanMessagePromptTemplate`: Represents the user's input.
- `SystemMessagePromptTemplate`: Represents instructions for the model, such as its role or behavior.
- `AIMessage`, `HumanMessage`, `SystemMessage`: Define different types of messages exchanged in the chat.
"""

# Initialize the Chat Model
chat_model = ChatOpenAI(model="gpt-3.5-turbo", temperature=0.7, openai_api_key=api_key)

"""
Step 2: Initialize the Chat Model
- `model`: Specifies the OpenAI model to use (e.g., `gpt-3.5-turbo`).
- `temperature`: Controls the randomness of the model's output (lower = more deterministic).
- `openai_api_key`: Passes the OpenAI API key to authenticate requests.
```

"""

# Define the Chat Prompt Template
system_message = SystemMessagePromptTemplate.from_template(
    "You are a helpful assistant that provides concise and accurate answers."
)
"""
Step 3a: Define the System Message
- This message provides context or instructions to the AI, guiding its behavior.
- Here, the AI is instructed to act as a helpful and concise assistant.
"""

human_message = HumanMessagePromptTemplate.from_template("{user_input}")
"""
Step 3b: Define the Human Message
- A placeholder `{user_input}` represents the user's input dynamically.
- It will later be replaced with actual text provided by the user.
"""

chat_prompt = ChatPromptTemplate.from_messages([system_message, human_message])
"""
Step 3c: Combine the Messages
- Combines `system_message` and `human_message` into a single structured prompt.
- This ensures the AI has instructions for the user's query.
"""

```python
# Prepare User Input
user_input = "Explain the difference between our solar system and one other well-known one"
```

"""
Step 4: Define the User Input
- The user's query is stored in a variable to dynamically populate the human message template.
"""

```python
# Format the Prompt
messages = chat_prompt.format_messages(user_input=user_input)
```

"""
Step 5: Format the Prompt
- Replaces `{user_input}` in the human message template with the actual user query.
- Creates a structured message format combining the system and human messages.
"""

```python
# Get Model Response
response = chat_model(messages)
```

"""
Step 6: Get the AI's Response
- Sends the formatted prompt to the model using the `chat_model` instance.
- Receives the AI's response, stored in the `response` variable.
"""

```python
# Display the AI Response
print("AI Response:", response.content)
```
"""

CHAPTER 8 TRANSFORMERS, GENERATIVE AI, AND LANGCHAIN

Step 7: Display the AI Response
- Extracts the generated text (from the `content` attribute of `response`).
- Prints the AI's response to the console for the user to review.
"""

>>> OUTPUT
AI Response: Our solar system consists of the Sun, eight planets (Mercury, Venus, Earth, Mars, Jupiter, Saturn, Uranus, Neptune), and various smaller celestial objects like moons, asteroids, and comets. It is located in the Milky Way galaxy.

One other well-known solar system is the TRAPPIST-1 system. TRAPPIST-1 is a star system located about 39 light-years away from Earth. It is known for having seven Earth-sized planets, three of which are located in the habitable zone where conditions might be right for liquid water to exist on the surface. This system is smaller and cooler than our Sun, and its planets are much closer to their star compared to the planets in our solar system.
"\nStep 7: Display the AI Response\n- Extracts the generated text (from the `content` attribute of `response`).\n- Prints the AI's response to the console for the user to review.\n"
""" We use gpt-3.5-turbo because it provides a balance between cost, speed, and quality. It is optimized for efficiency, making it ideal for real-time applications and widely adopted tasks that require robust yet economical solutions.

Other popular OpenAI models include gpt-4 for advanced reasoning, text-davinci-003 for high-quality outputs, and smaller models like curie, babbage, and ada for simpler tasks. Each model is tailored for specific needs, from cost-saving to handling complex scenarios. """

8.4.1 Using LLMs with LangChain

Developers can leverage LangChain with LLMs and Chat Models to customize many factors like prompts, integrate memory, and link peripheral tools to augment the model capabilities. This section explores it hands-on to understand key steps and techniques.

Listing 8-5 is a simple demo program leveraging LangChain with LLMs.

Listing 8-5. Generating Success Tips with OpenAI's LLM

```
"""This program uses OpenAI's language model to answer the
question "How can I make my day successful?""""
# Read the openai api key from your text filE
f = open('C:\\Users\\Shailendra Kadre\\Desktop\\OPEN_AI_
KEY.txt')
api_key = f.read()
# To avoid depricated method warnings, install the following if
n ot done already
!pip install -U langchain-openai
from langchain_openai import OpenAI

# Initialize the LLM
my_llm = OpenAI(openai_api_key=api_key)
print(my_llm('How can I make my day successful?'))
OUTPUT 1
1. Start your day with a positive attitude: A positive
mindset can set the tone for a successful day. Practice
gratitude, affirmations, or meditation to start your day on a
positive note.
2. Set realistic goals: Identify what you want to achieve
for the day and set realistic goals. This will help you stay
focused and motivated throughout the day.
```

3. Plan your day: Take a few minutes to plan out your day, including tasks and appointments. This will help you stay organized and prioritize your tasks.
4. Take breaks: It's important to take breaks throughout the day to recharge and avoid burnout. Use your breaks to relax, stretch, or do something you enjoy.
5. Stay hydrated and nourished: Make sure to drink enough water and eat healthy meals and snacks throughout the day. This will help you stay energized and focused.
6. Stay organized and declutter: A cluttered workspace can lead to a cluttered mind. Take a few minutes to declutter and organize your workspace to improve your focus and productivity.
7. Prioritize tasks: Identify the most important tasks and work on them first. This will help you avoid feeling overwhelmed and ensure that the most critical tasks are completed.
8. Focus on one task at a time: Mult

```
# Use generate to get the more details
result = my_llm.generate(['How can I make my day successful?'])
# Print only the output text
print("Output Text:", result.generations[0][0].text.strip())
```

OUTPUT 2

Output Text: 1. Set goals: Start your day by setting realistic and achievable goals. This will help you stay focused and motivated throughout the day.
2. Wake up early: Waking up early gives you more time to accomplish your tasks and sets a positive tone for the day.
3. Exercise: Start your day with some form of physical activity. This will boost your energy levels and improve your overall mood.

4. Eat a healthy breakfast: A nutritious breakfast will provide you with the energy and nutrients you need to stay productive throughout the day.
5. Prioritize tasks: Make a to-do list and prioritize your tasks based on their importance. This will help you stay organized and ensure that you complete the most important tasks first.
6. Take breaks: It's important to take breaks throughout the day to avoid burnout. Use your breaks to relax, stretch, or do something you enjoy.
7. Stay hydrated: Drinking enough water throughout the day will help you stay focused and energized.
8. Stay positive: A positive mindset can make a huge difference in your productivity and overall well-being. Focus on the good things and try to stay optimistic.
9. Limit distractions: Minimize distractions such as social media, emails, or phone calls during work hours.

8.4.2 Using Chat Models with LLMs

Let's discuss its background theory and description aspects of LangChain with OpenAI chat models. Let's first consider the following terms.

- **ChatPromptTemplate** is used for creating an end-to-end complete conversation prompt. ChatPromptTemplate combines various message types to make sure the conversation is structured properly for the LLM. For example, it can contain a system instruction like "You are a helpful assistant" and also have a user input like "How do I prepare Paneer Butter Masala, a popular Indian dish?"

- **HumanMessagePromptTemplate** outlines the user's input in a two-way conversation. For example, for a user query, "What is the commercial capital of the US?" this template captures the given query and then passes it on to the LLM like OpenAI Chatbot. HumanMessagePromptTemplate facilitates user interactions to be dynamic and easy to handle.

- **SystemMessagePromptTemplate** is responsible for setting the rules (commands) for the AI. For instance, the role or behavior of AI. Let's construct this instruction to AI, "You are a popular math professor who makes even complex concepts as easy to understand." Such instruction(s) prompt the LLM to answer in an explicit style. It also makes sure the AI remains on track (task) all through the chat.

These three templates combined together to structure conversations with LLMs. The following sections look into the usage of these templates.

- **SystemMessagePromptTemplate** sets the AI's role or behavior.

- **HumanMessagePromptTemplate** captures the user's input.

- **ChatPromptTemplate** combines these messages to create a full conversation for the AI to process.

Now, it is time to look into various available (popular) OpenAI Models for our use. Each one is created with a specific purpose in mind. For instance, use `gpt-3.5-turbo` to create a chatbot or `davinci` for writing essays. The following are the more popular ones.

- **gpt-4** is best for complex reasoning tasks.

- **gpt-3.5-turbo** is efficient and cost-effective for most applications.

- **text-davinci-003** is great for creative writing or summarization.

- **text-curie-001** is faster and cheaper and ideal for simple tasks.

Building a Friendly Chat Assistant with LangChain and GPT-3.5-Turbo

Listing 8-6 is demo code covering the concepts covered in this section. This program uses LangChain's templates to structure a conversation with GPT-3.5-Turbo. The SystemMessagePromptTemplate sets the AI's role as a friendly assistant, while the HumanMessagePromptTemplate dynamically captures user input. These are combined using a ChatPromptTemplate to create a full prompt. The AI processes this prompt and generates a clear and concise response based on the user's query.

Listing 8-6. Building a Friendly Chat Assistant with LangChain and GPT-3.5-Turbo

```
f = open('C:\\Users\\Shailendra Kadre\\Desktop\\OPEN_AI_KEY.txt')
api_key = f.read()
from langchain.chat_models import ChatOpenAI
from langchain.prompts.chat import (
    ChatPromptTemplate,
    HumanMessagePromptTemplate,
    SystemMessagePromptTemplate,
)
```

```python
# Initialize the Chat Model
chat_model = ChatOpenAI(model="gpt-3.5-turbo", temperature=0.7,
openai_api_key=api_key)

# Define the System Message Template
system_message = SystemMessagePromptTemplate.from_template(
    "You are a friendly assistant that answers questions
    clearly and concisely."
)

# Define the Human Message Template
human_message = HumanMessagePromptTemplate.from_template("{user_input}")

# Combine into a Chat Prompt Template
chat_prompt = ChatPromptTemplate.from_messages([system_message,
human_message])

# Example User Input
user_input = "What are some fun weekend activities for kids?"

# Format the Prompt
messages = chat_prompt.format_messages(user_input=user_input)

# Get the Model Response
response = chat_model(messages)

# Display the AI Response
print("AI Response:", response.content)
```

AI Response: Some fun weekend activities for kids include going to the park, visiting a zoo or aquarium, having a picnic, going on a nature hike, going to a children's museum, having a movie or game night at home, or doing a craft or science project together.

CHAPTER 8   TRANSFORMERS, GENERATIVE AI, AND LANGCHAIN

Let's now turn to a couple of new concepts.

- **Document loaders** are deployed to read and load data from files like PDFs, Word documents, or web pages. Document loaders reformat the content into a form that works with AI models. For example, we can upload a Microsoft Word document and use it for question answering applications.

- **Document integration loaders** integrate data from many documents to create a single document. This way, large sets of information can be easily handled in one place. For instance, Document Integration Loaders can combine multiple files from a folder to generate a single knowledge base for the AI.

- **Document transformers** process, clean, and maybe reformat the loaded content to make it work for the AI. For example, document transformers can be used to clean the raw input text, eliminate superfluous parts, or even summarize bigger documents. This way helps the AI models to work more proficiently with the data.

Business Data Preparation for AI Applications

Listing 8-7 examines the demo code for the document and data preparation functions. This program reads data from a PDF and a text file. It combines the data into one set for easier handling. It processes and cleans the data by splitting it into smaller chunks. Finally, the cleaned data is ready for AI to use, such as answering questions. This program demonstrates how to prepare data for AI use.

- Document Loade: PDF and text files are read using `PyPDFLoader` and `TextLoader`.

CHAPTER 8　TRANSFORMERS, GENERATIVE AI, AND LANGCHAIN

- Document Integration: All loaded files are combined into one list **for** easy management.

- Document transformers: Data is cleaned and split into manageable chunks **for** AI models.

Listing 8-7. Business Data Preparation for AI Applications

```
# Install it if not done already.
#!pip install pypdf
# Importing necessary libraries
from langchain.document_loaders import PyPDFLoader  # To load and read PDF files
from langchain.document_loaders import TextLoader  # To load and read text files
from langchain.text_splitter import CharacterTextSplitter  # To split text into smaller chunks

# Define a simple text-cleaning function
def clean_text(text):
    """Remove unnecessary characters and extra spaces."""
    import re
    text = re.sub(r"\s+", " ", text)  # Replace multiple spaces with a single space
    text = re.sub(r"[^a-zA-Z0-9.,!? ]", "", text)  # Remove special characters
    return text.strip()

# Step 1: Load documents
pdf_loader = PyPDFLoader("business_report.pdf")  # Replace with your business PDF file
pdf_documents = pdf_loader.load()  # Load and parse the PDF content
```

```python
text_loader = TextLoader("meeting_notes.txt")  # Replace with your business text file
text_documents = text_loader.load()  # Load and parse the text file content

# Step 2: Integrate documents
all_documents = pdf_documents + text_documents

# Step 3: Transform documents
splitter = CharacterTextSplitter(chunk_size=500, chunk_overlap=50)
transformed_docs = []
for doc in all_documents:
    chunks = splitter.split_text(doc.page_content)  # Use 'page_content' instead of 'content'
    cleaned_chunks = [clean_text(chunk) for chunk in chunks]  # Clean each chunk
    transformed_docs.extend(cleaned_chunks)

# Step 4: Use the transformed data for AI (Example: Question Answering)
print("Transformed and cleaned document chunks ready for AI processing:")
for doc in transformed_docs[:5]:  # Print a few examples
    print(doc)
```
OUTPUT:
Transformed and cleaned document chunks ready for AI processing:
Adobe Acrobat PDF Files Adobe Portable Document Format PDF is a universal file format that preserves all of the fonts, formatting, colours and graphics of any source document, regardless of the application and platform used to create it.

CHAPTER 8   TRANSFORMERS, GENERATIVE AI, AND LANGCHAIN

```
............ Compact PDF files are smaller than their source
files and download a page at a time for fast display on the
Web. Step 2 Integrate documents Combine the loaded PDF and text
documents into a single list.
```

Let's go over a few more useful concepts and write a demo program demonstrating their use as usual.

- **Text embeddings** convert words or sentences into numeric vectors. We do so as machines (algorithms) need data in number format. For instance, "car" and "vehicle" may have similar embeddings or vectors since they are related. AI models consume embeddings to compare and analyze text.

- A **vector store** saves and categorizes embeddings so that AI models can search for related meanings. For example, a vector store saves the embedding of "car" to find similar terms like "truck" or "automobile." Vector stores make searching text more efficient.

- A **vector store retriever** discovers similar (matching) vectors from the vector store. A vector store retriever helps to search for answers (or related text) swiftly. For instance, if you search for the word "train," it recovers embeddings for "railway" or "locomotive." This process speeds up searches in AI applications. Next, we look at a demo program using these concepts.

CHAPTER 8   TRANSFORMERS, GENERATIVE AI, AND LANGCHAIN

Text Embedding Search with Vector Store and Retriever

The program shown in Listing 8-8 creates text embeddings to represent words as numbers. It stores the embeddings in a vector store for easy access. Using a vector store retriever, it finds related words based on their meanings. This demonstrates how AI can understand and compare text efficiently.

- **Text embeddings**: Words like "car" and "bicycle" are turned into numbers using a model.

- **Vector store**: These numbers are saved for future searches.

- **Vector store retriever**: Finds the closest matches to a query like "automobile."

- **Purpose**: Shows how AI connects related words using embeddings and vector search.

Listing 8-8. Text Embedding Search with Vector Store and Retriever

```
# Install the library needed to turn text into numbers
(embeddings) for the program.
# !pip install sentence-transformers
# FAISS (Facebook AI Similarity Search) is a library for
efficient similarity search and clustering of dense vectors.
# Install the FAISS library optimized for CPU usage
# !pip install faiss-cpu
# Suppress a specific warning when using Hugging Face models in
environments that don't support symlinks (like on Windows).
import os
os.environ["HF_HUB_DISABLE_SYMLINKS_WARNING"] = "1"
# Import necessary libraries
```

CHAPTER 8 TRANSFORMERS, GENERATIVE AI, AND LANGCHAIN

```python
from sentence_transformers import SentenceTransformer
# For creating text embeddings
from langchain.vectorstores import FAISS   # To create and manage a vector store
from langchain.schema import Document  # For creating Document objects
from langchain.embeddings import HuggingFaceEmbeddings
# Use HuggingFace for embeddings

# Step 1: Create a text embedding model using HuggingFaceEmbeddings
embedding_model = HuggingFaceEmbeddings(model_name="sentence-transformers/all-MiniLM-L6-v2")

# Sample text data
texts = ["car", "vehicle", "train", "bicycle", "airplane"]
# Words to embed and compare

# Step 2: Convert text to Document objects
documents = [Document(page_content=text) for text in texts]
# Wrap texts in Document objects

# Step 3: Create a vector store using HuggingFaceEmbeddings
vector_store = FAISS.from_documents(documents, embedding_model)  # Create vector store with embeddings

# Step 4: Use a vector store retriever
query = "automobile"  # Input to find related terms
query_embedding = embedding_model.embed_documents([query])
# Create embedding for the query

# Step 5: Use the retriever to get relevant documents using `invoke` (not deprecated)
```

```
retriever = vector_store.as_retriever()  # Initialize retriever
from the vector store
results = retriever.invoke(query)  # Use invoke to find matches
(replaces get_relevant_documents)

# Step 6: Display results
print(f"Query: {query}")
print("Top matches:")
for result in results:
    print(result.page_content)  # Print the matching
document content
OUTPUT:
Query: automobile
Top matches:
car
vehicle
bicycle
airplane
```

Multi-query retrievers advance search accuracy by creating multiple query variations of the input, while context compression summarizes input to handle long contexts efficiently. The following are brief descriptions.

- **Multi-query retriever** helps improve search precision by creating diverse versions of a query. For instance, if the user input query is "Best books," Multi-Query Retriever also produces associated versions like "Top books to read" or "Popular books." This process aids in getting more complete search results. It's like requesting the same question in multiple ways to get the best answer from LLMs.

- **Context compression** condenses long input texts so that the focus remains on the most important parts. For example, for a ten-page input document, the context compression component can compress it into a few key paragraphs. This comes in handy, especially when a full document can't be because of size limitations. It's analogous to writing a summary of a lengthy story to save time.

Next, let's analyze a demo program showing the usage of multi-query retriever and context compression.

Smart Query Helper with LangChain and OpenAI

The program shown in Listing 8-9 takes a main question and splits it into smaller questions to find better answers using a multi-query retriever. It also combines and compresses the results into a short and clear summary. It uses LangChain and OpenAI for building the helper. The goal is to make searching smarter and faster!

- Save some example text in a file named sample_text.txt.
- Run this program to ask a question about the text.
- The program retrieves and summarizes relevant information along with sources!

Listing 8-9. Smart Query Helper with LangChain and OpenAI

```
# Read the open ai api key from your text filr
f = open('C:\\Users\\Shailendra Kadre\\Desktop\\OPEN_AI_KEY.txt')
api_key = f.read()
from langchain.chains import RetrievalQAWithSourcesChain
from langchain.chat_models import ChatOpenAI
```

CHAPTER 8 TRANSFORMERS, GENERATIVE AI, AND LANGCHAIN

```python
from langchain.vectorstores import FAISS
from langchain.embeddings.openai import OpenAIEmbeddings
# Import OpenAI's embeddings
from langchain.text_splitter import RecursiveCharacterTextSplitter
from langchain.document_loaders import TextLoader

# Step 1: Load some example text data
loader = TextLoader('sample_text.txt')
documents = loader.load()

# Step 2: Split the text into manageable chunks
text_splitter = RecursiveCharacterTextSplitter(chunk_size=500, chunk_overlap=50)
docs = text_splitter.split_documents(documents)

# Step 3: Use FAISS as the retriever with OpenAI embeddings
embeddings = OpenAIEmbeddings(openai_api_key=api_key)  # Pass the API key here
retriever = FAISS.from_documents(docs, embeddings).as_retriever()

# Step 4: Set up the ChatOpenAI model with the API key
llm = ChatOpenAI(temperature=0, openai_api_key=api_key)  # Pass the API key here

# Step 5: Combine the retriever and the model
qa_chain = RetrievalQAWithSourcesChain.from_chain_type(
    llm=llm, retriever=retriever, return_source_documents=True
)

# Step 6: Ask a complex question and get a smart answer
# Pass the query as a dictionary with the correct key: 'question'
```

```
query = {"question": "How does AI help teachers save time in
their daily tasks?"}
result = qa_chain(query)   # Use the correct input format

# Display the compressed answer and sources
print("Summary Answer:", result['answer'])
print("\nSources:", result['sources'])
```
Summary Answer: AI helps teachers save time in their daily
tasks by automating repetitive tasks such as grading
assignments and creating quizzes, allowing educators to focus
more on teaching. AI chatbots can also assist students by
providing instant feedback and answering questions, making
learning more interactive and engaging.

Sources: sample_text.txt

8.4.3 Working with Prompt Templates and Few-Shot Templates

Prompt templates and few-shot templates are indispensable tools that guide AI models to produce precise responses. Both these templates offer a structured method to frame inputs. Prompt templates and few-shot templates improve the model's comprehension of your exact requirements. The following describes the use of these templates.

- **Prompt templates** are pre-designed formats to ask questions or give instructions to your AI model. They ensure the given input is clear and organized; for example, *"Interpret the text to German: {text}"*. In this example, *{text}* is simply a placeholder where you can add any input. This arrangement supports your AI model answer accurately.

- **Few-shot templates** give examples to the AI before asking a question. This shows the AI how to respond. For instance, *Question: What is generative AI? Answer: Generative AI is a smart computer system that can generate text and images. Now your turn: Question: {question}"*. In this example, *{question}* is a placeholder where you can input your own question. The AI uses this type of construct to give better answers.

Next, let's look at a demo program to show the usage of prompt templates and few-shot templates. AI model's responses with and without templates are also compared with comments.

Guiding AI Responses with and Without Templates

The following is a Python demo using LangChain to show the difference between using no templates and using prompt templates.

```
# Read the open ai api key from your text filr
f = open('C:\\Users\\Shailendra Kadre\\Desktop\\OPEN_AI_KEY.txt')
api_key = f.read()
```

Without Templates

```
# Import the ChatOpenAI class for interacting with OpenAI's
language models
from langchain.chat_models import ChatOpenAI

# Initialize the OpenAI model with default settings
llm = ChatOpenAI(temperature=0, openai_api_key=api_key)

# Ask the AI a simple question without using a template
response = llm.predict("Explain what Generative AI is.")
print(response)  # Print the AI's response
```

OUTPUT:
Generative AI refers to a type of artificial intelligence that is capable of creating new content, such as images, text, or music, that is original and not based on existing data. This type of AI uses algorithms to generate new content by learning patterns and structures from a dataset and then creating new content based on those patterns. Generative AI can be used in a variety of applications, such as creating realistic images, generating text for chatbots, or composing music. It has the potential to revolutionize creative industries by automating the process of content creation and enabling new forms of artistic expression.

- Response without template: The answer is longer, detailed, and covers multiple aspects.
- Issue: The AI is freeform, which may result in verbose or inconsistent responses. The tone and content might vary depending on the question.

With Prompt Template

```
# Import necessary components
from langchain.prompts import PromptTemplate  # For creating structured prompts
from langchain.chat_models import ChatOpenAI  # For interacting with OpenAI's language models
# Define a prompt template with placeholders
template = PromptTemplate(
    input_variables=["topic"],  # Define the placeholder 'topic'
    template="Explain what {topic} is in simple terms.",
    # Structure the input
)
```

```
# Initialize the OpenAI model with default settings
llm = ChatOpenAI(temperature=0, openai_api_key=api_key)

# Fill in the placeholder with a specific topic
prompt = template.format(topic="Generative AI")  # Replace
'topic' with "Generative AI"

# Get the AI's response using the formatted prompt
response = llm.predict(prompt)
print(response)  # Print the AI's response
OUTPUT:
```
Generative AI is a type of artificial intelligence that can create new content, such as images, text, or music, based on patterns it has learned from existing data. It uses algorithms to generate new, original content that is similar to what it has been trained on. This technology can be used in a variety of applications, such as creating realistic images, generating personalized recommendations, or even composing music.

With Few-Shot Template

Listing 8-10 depicts the program for working with a few-shot template.

Listing 8-10. Code demo for working with a ew-shot template

```
# Import necessary components
from langchain.prompts import FewShotPromptTemplate,
PromptTemplate  # For few-shot and regular prompts
from langchain.chat_models import ChatOpenAI  # For interacting
with OpenAI's language models
# Define examples to show the AI how to respond
examples = [
    {"question": "What is Machine Learning?", "answer":
    "Machine Learning is a system where computers learn from
    data."},
```

```
    {"question": "What is AI?", "answer": "AI is a field of
    computer science that makes machines smart."},
]

# Create a PromptTemplate for each example
example_prompt = PromptTemplate(
    input_variables=["question", "answer"],  # Define the
    placeholders
    template="Question: {question}\nAnswer: {answer}",
    # Format for each example
)

# Define the prefix and suffix for the few-shot template
prefix = "You are an expert in technology. Answer the following
questions in simple terms:"
suffix = "Now your turn: Question: {question}"

# Create a FewShotPromptTemplate
few_shot_template = FewShotPromptTemplate(
    examples=examples,  # Provide the examples
    example_prompt=example_prompt,  # Use the PromptTemplate
    for examples
    prefix=prefix,  # Add the prefix text
    suffix=suffix,  # Add the suffix text
    input_variables=["question"],  # Define the placeholder for
    the final question
)

# Initialize the OpenAI model
llm = ChatOpenAI(temperature=0, openai_api_key=api_key)
# Provide your API key
```

```
# Fill in the placeholder with your question
prompt = few_shot_template.format(question="What is Generative
AI?")   # Replace 'question' with your query

# Get the AI's response using the few-shot template
response = llm.predict(prompt)
print(response)   # Print the AI's response
OUTPUT:
```
Answer: Generative AI is a type of artificial intelligence that can create new content, such as images, music, or text, based on patterns it has learned from existing data.

- **Response with the template**: The answer is concise and aligns with the examples provided.
- **Benefit**: The examples serve as a guide, instructing the AI on how to respond in a specific style, resulting in output that is focused, consistent, and simpler.

8.4.4 Parsing and Serialization

When LangChain and LLMs are in the background, we use parsing to structure and process user input queries. This way, AI models can better understand user queries. In input parsing, we break the user queries into more manageable pieces, which make it easier for LLMs from OpenAI or others to process. Input parsing may include tasks such as converting the input text to lowercase, splitting it into tokens (words), or removing unwanted text or other undesired information.

There exists output parsing that is employed to extract relevant and usable information from the LLM response. Output parsing ensures the easy understanding and consumption of LLM responses down the line. One useful example of output parsing can be summarization and extraction of only the key parts of the LLM's response. Next, let's examine demo programs to illustrate these concepts.

Input and Output Parsing in LangChain and OpenAI

The program shown in Listing 8-11 demonstrates how to use Pydantic, Datetime, CommaSeparatedList, and OutputFixing parsers in LangChain. Each section validates or corrects input data and outputs the results clearly and understandably.

The program parses with and without the LLM.

Listing 8-11. Input and Output Parsing in LangChain and OpenAI

```
# Read the open ai API key from your text file
f = open('C:\\Users\\Shailendra Kadre\\Desktop\\OPEN_AI_
KEY.txt')
api_key = f.read()
# Import necessary parsers from LangChain
from langchain.output_parsers import PydanticOutputParser
# For parsing data into Pydantic models
from langchain.output_parsers import DatetimeOutputParser
# For parsing datetime strings
from langchain.output_parsers import
CommaSeparatedListOutputParser  # For parsing comma-
separated lists
from langchain.output_parsers import OutputFixingParser
# For fixing and handling parser errors
from pydantic import BaseModel  # Import Pydantic to define
structured data models
# Sample input data
input_data = "2025-01-22 12:30:00, Shailendra, Engineering,
Deep Learning"  # Example input string
```

```python
# Define a Pydantic model for structured data
class DataModel(BaseModel):
    date_time: str
    name: str
    field: str
    subject: str
# First, use DatetimeOutputParser to parse the date and time
# from input data
from langchain.output_parsers import DatetimeOutputParser
from datetime import datetime

# Example input data
input_data = "2025-01-22 12:30:00, Shailendra, Engineering, Deep Learning"

# Extract the datetime string (first part before the comma)
datetime_str = input_data.split(',')[0].strip()

# Convert the datetime string to ISO 8601 format (with
# milliseconds and 'Z' suffix)
try:
    iso_datetime_str = datetime.strptime(datetime_str, "%Y-%m-%d %H:%M:%S").isoformat(timespec='milliseconds') + "Z"
    print("ISO 8601 formatted datetime:", iso_datetime_str)
    # Output the formatted datetime string
except ValueError as e:
    print("Datetime Parsing Error:", e)
    iso_datetime_str = None  # Handle the error gracefully

# Initialize the DatetimeOutputParser if datetime_str is valid
if iso_datetime_str:
    datetime_parser = DatetimeOutputParser()  # Initialize the datetime parser
```

```
    try:
        # Now parse with the new parser if iso_datetime_str
        is valid
        datetime_parsed = datetime_parser.parse(iso_datetime_
        str)  # Pass the correctly formatted datetime string
        print("Parsed Date and Time:", datetime_parsed)
        # Output the parsed datetime
    except Exception as e:
        print("Parsing Error:", e)
else:
    print("Invalid datetime format.")
ISO 8601 formatted datetime: 2025-01-22T12:30:00.000Z
Parsed Date and Time: 2025-01-22 12:30:00
# Next, use CommaSeparatedListOutputParser to parse the
remaining comma-separated values
comma_str = ', '.join(input_data.split(',')[1:])  # Extract the
part after the datetime and join into a string
comma_parser = CommaSeparatedListOutputParser()  # Initialize
the parser for comma-separated lists
comma_parsed = comma_parser.parse(comma_str)  # Parse the
comma-separated values
print("Parsed Comma-Separated List:", comma_parsed)  # Output
the parsed list
Parsed Comma-Separated List: ['Shailendra', 'Engineering',
'Deep Learning']
# Let's also demonstrate OutputFixingParser for fixing and
parsing faulty inputs
from langchain.llms import OpenAI
from langchain.llms import OpenAI
from langchain.output_parsers import DatetimeOutputParser,
OutputFixingParser
from datetime import datetime
```

CHAPTER 8 TRANSFORMERS, GENERATIVE AI, AND LANGCHAIN

```python
# Replace with your actual OpenAI API key
api_key = api_key  # <-- Ensure this is your actual API key

# Initialize the LLM with the correct API key
llm = OpenAI(temperature=0.7, openai_api_key=api_key)

# Example input data with datetime in the correct format
datetime_str = "2025-01-22T12:30:00.000Z"  # Ensure this matches the required format

# Initialize the datetime parser
datetime_parser = DatetimeOutputParser()

# Use the OutputFixingParser with the LLM
output_fixer = OutputFixingParser.from_llm(parser=datetime_parser, llm=llm)

# Fix and parse the datetime string
try:
    fixed_output = output_fixer.parse(datetime_str)  # Parsing the datetime string
    print("Fixed Output:", fixed_output)  # Output the fixed parsed result
except Exception as e:
    print("Error:", e)
Fixed Output: 2025-01-22 12:30:00
# Demo of pydantic parser
# Install the library if not done already
#!pip install pydantic
from langchain.output_parsers import PydanticOutputParser
class Planet(BaseModel):
    name: str = Field(description="Name of a planet")
    discoveries: list = Field(description="Python list of three facts about it")
```

```
query = 'Name a well known planet and a list of three facts
about it'
parser = PydanticOutputParser(pydantic_object=Planet)
print(parser.get_format_instructions())
```
The output should be formatted as a JSON instance that conforms to the JSON schema below.

As an example, for the schema {"properties": {"foo": {"title": "Foo", "description": "a list of strings", "type": "array", "items": {"type": "string"}}}, "required": ["foo"]}
the object {"foo": ["bar", "baz"]} is a well-formatted instance of the schema. The object {"properties": {"foo": ["bar", "baz"]}} is not well-formatted.

Here is the output schema:
```
{"properties": {"name": {"description": "Name of a planet", "title": "Name", "type": "string"}, "discoveries": {"description": "Python list of three facts about it", "items": {}, "title": "Discoveries", "type": "array"}}, "required": ["name", "discoveries"]}
```

```
from langchain.prompts import PromptTemplate
from langchain.llms import OpenAI  # Ensure that OpenAI is correctly imported

# Initialize your OpenAI model with the correct API key
#api_key = "your_openai_api_key_here"  # Replace with your actual API key
llm = OpenAI(temperature=0.7, openai_api_key=api_key)
```

```python
# Assuming you're working with simple text output
# Create a prompt template with placeholders for 'query' and
'format_instructions'
prompt = PromptTemplate(
    template="Answer the user query.\n{format_instructions}\
    n{query}\n",
    input_variables=["query"],
    partial_variables={"format_instructions": parser.get_
    format_instructions()},  # Adjust if needed
)

# Format the prompt with the user's query
_input = prompt.format_prompt(query='Name a well known planet
and a list of three facts about it')

# Send the formatted prompt to the model and receive the output
output = llm(_input.to_string())

# Output is plain text, so you can directly print or process it
print("Model Output:", output)
Model Output:
{"name": "Earth", "discoveries": ["Earth is the third planet
from the Sun", "It is the only known planet to have liquid
water on its surface", "Earth has a single natural satellite,
the Moon"]}
```

Serialization in the context of AI means we can save (in a format suitable for storage) the AI's output, transmit it, and retrieve it later when required. During this process, we can even convert data to JSON or other supported structures. Serialization techniques advance the way AI interacts with users and handles data. The serialization process thus makes the entire process more efficient, reliable, and effective.

Input and Output with Serialization in LangChain and OpenAI

Listing 8-12 shows how to save and reload a setup for a language model (an AI that understands and generates text). It uses a tool called LangChain and OpenAI's API to do the following.

1. Create a task where the AI translates English text into French.

2. Save the setup (like a recipe) into a file so it can be reused later.

3. Load the saved setup back from the file.

4. Test the loaded setup by asking the AI to translate a sentence.

It's like writing down a recipe, storing it, and then using it again later to cook the same dish!

Listing 8-12. Input and Output with Serialization in LangChain and OpenAI

```
# Read the open ai API key from your text file
f = open('C:\\Users\\Shailendra Kadre\\Desktop\\OPEN_AI_
KEY.txt')
api_key = f.read()
import json
from langchain.prompts import PromptTemplate
from langchain_openai import ChatOpenAI
from langchain.schema.runnable import RunnableSequence

# Step 1: Provide your OpenAI API key
#api_key = "your_openai_api_key_here"  # Replace with your
actual OpenAI API key
```

CHAPTER 8 TRANSFORMERS, GENERATIVE AI, AND LANGCHAIN

```python
# Step 2: Initialize the ChatOpenAI model
llm = ChatOpenAI(model="gpt-3.5-turbo", temperature=0, openai_api_key=api_key)

# Step 3: Create a prompt template
prompt = PromptTemplate(
    input_variables=["text"],
    template="Translate this to French: {text}"
)

# Step 4: Create a RunnableSequence (replaces LLMChain)
chain = prompt | llm

# Step 5: Serialize the RunnableSequence manually
serialization_path = "llm_chain.json"
# Save the necessary configuration parameters for reconstruction
chain_dict = {
    "llm": {
        "model": "gpt-3.5-turbo",
        "temperature": 0,
        "openai_api_key": api_key
    },
    "prompt": {
        "template": prompt.template,
        "input_variables": prompt.input_variables
    }
}
with open(serialization_path, "w") as f:
    json.dump(chain_dict, f)
print("Serialized RunnableSequence saved to llm_chain.json.")
```

```
# Step 6: Load the serialized data
with open(serialization_path, "r") as f:
    chain_data = json.load(f)

# Step 7: Reconstruct the RunnableSequence manually
# Recreate the ChatOpenAI instance
loaded_llm = ChatOpenAI(
    model=chain_data["llm"]["model"],
    temperature=chain_data["llm"]["temperature"],
    openai_api_key=chain_data["llm"]["openai_api_key"]
)

# Recreate the PromptTemplate instance
loaded_prompt = PromptTemplate(
    input_variables=chain_data["prompt"]["input_variables"],
    template=chain_data["prompt"]["template"]
)

# Recreate the RunnableSequence
loaded_chain = loaded_prompt | loaded_llm
print("Loaded RunnableSequence successfully!")

# Step 8: Use the loaded chain to process input
input_text = {"text": "Hello, how are you?"}
output = loaded_chain.invoke(input_text)

# Extract only the main content from the output
if isinstance(output, dict) and "content" in output:
    output_text = output["content"]
else:
    output_text = output.content if hasattr(output, "content")
    else str(output)
```

```
# Step 9: Print concise output
print(f"Output: {output_text}")
Serialized RunnableSequence saved to llm_chain.json.
Loaded RunnableSequence successfully!
Output: Bonjour, comment vas-tu ?
```

8.4.5 Customizing Model Inputs and Handling Outputs

Customizing inputs and handling outputs goes much beyond simple parsing. While the focus of the LLM parsing process is to break inputs and outputs (to and from LLMs) into structured formats, customization focuses on adapting inputs and outputs to meet precise goals. Customization of LLM inputs involves shaping prompts for clarity, relevance, and context. This involves adding comprehensive instructions and utilizing examples to guide the model, or it can involve formatting text for improved understanding. For example, while processing a user query to LLM for summarizing a piece of text, the customization process can include key points, or it can specify word limits in the prompt itself to ensure more accurate results. Customization goes much beyond parsing, which is discussed in the previous section. The precise goal of customizing inputs and outputs for LLMs is to craft them to guide the model efficiently toward the anticipated result with the desired information in the desired format. For instance, LLM-driven chatbots may need trimming of output to fit the designed character limits. The customization process, in this case, may even rephrase the output to maintain the desired consistency in tone. Similarly, applications like report generators often require precise formatting of outputs according to structured templates. The input/output customization process makes sure that the LLM responses are usable and aligned with the planned purpose.

CHAPTER 8 TRANSFORMERS, GENERATIVE AI, AND LANGCHAIN

Customization and handling often require iterative strategies, where inputs are tested and refined based on the output quality. Feedback loops are created to find gaps and improve performance. Thoughtful customization of inputs and outputs for LLMs can help maximize efficiency for tasks such as text summarization, social media sentiment analysis, and customer service. Customization is a comprehensive approach that brings accuracy, relevance, and efficiency in leveraging LLMs for business and other real-life applications.

Input and Output Customization with LangChain and OpenAI

The following describes how the Listing 8-13 program works.

1. **Imports required libraries.** Loads necessary libraries from LangChain to interact with OpenAI's model.

2. **Defines a PromptTemplate.** Sets up a prompt template to instruct the model on how to summarize a given topic in a specified word limit.

3. **Sets up API key.** Prepares the API key for OpenAI access.

4. **Creates a LLM (language model) instance.** Initializes OpenAI's GPT-3.5-Turbo model with specific settings like temperature and max tokens.

5. **Inputs data.** Defines the `topic` and `word_limit` for the summary.

6. **Formats the prompt.** Uses the `PromptTemplate` to insert the `topic` and `word_limit` into the prompt.

7. **Generates a response.** Sends the formatted prompt to the GPT-3.5 model and receives a response.

8. **Refines the output.** Processes the model's response by breaking it down into key points for improved readability.

9. **Displays the output.** Prints both the formatted input and the refined output in bullet points.

Listing 8-13. Input and Output Customization with LangChain and OpenAI

```
# Read the open ai API key from your text file
f = open('C:\\Users\\Shailendra Kadre\\Desktop\\OPEN_AI_KEY.txt')
api_key = f.read()
# Import required libraries
from langchain_openai import ChatOpenAI  # Updated import for
the OpenAI model
from langchain.prompts import PromptTemplate

# Step 1: Define the input customization using a detailed
prompt template
# This prompt provides clear instructions and a specific
context for the LLM.
prompt_template = PromptTemplate(
    input_variables=["topic", "word_limit"],
    template="""
You are an expert content writer. Write a summary about
"{topic}" in {word_limit} words.
Focus on the key points, avoid unnecessary details, and ensure
readability for a general audience.
"""
)
```

```python
# Step 2: Provide your OpenAI API key
# Replace YOUR_API_KEY_HERE with your actual OpenAI API key.
api_key = api_key

# Step 3: Create an instance of OpenAI's GPT-3.5-Turbo model
# using LangChain
llm = ChatOpenAI(
    model="gpt-3.5-turbo",  # Specify the turbo model
    temperature=0.7,  # Controls creativity
    max_tokens=300,  # Maximum response length
    openai_api_key=api_key  # Pass your API key
)

# Step 4: Provide input values for the topic and word limit
input_data = {
    "topic": "The impact of climate change on global
    agriculture",
    "word_limit": 50
}
# Step 5: Format the prompt using the input values
formatted_input = prompt_template.format(**input_data)
# Format the prompt with input_data

# Step 6: Execute the model using the formatted input
response = llm(formatted_input)

# Step 7: Handle and customize the output
# Access the content of the AIMessage object using .content
raw_output = response.content  # Correct way to access the
response content

# Post-process the output to extract and format key points.
def refine_output(raw_output):
    """
```

```
    This function extracts key points and formats the output
    for better readability.
    """
    key_points = [point.strip() for point in raw_output.
    split(".") if point.strip()]
    return "\n".join(f"- {point}." for point in key_points)

# Refine the raw output
refined_output = refine_output(raw_output)

# Step 8: Display the input and final output
print("Customized Input:")
print(formatted_input)  # Print the final customized input
print("\nRefined Output:")
print(refined_output)  # Display the refined and formatted
response
PROGRAM OUTPUT:
Customized Input:
```

You are an expert content writer. Write a summary about "The impact of climate change on global agriculture" in 50 words. Focus on the key points, avoid unnecessary details, and ensure readability for a general audience.

Refined Output:
- Climate change has profound effects on global agriculture, leading to shifts in growing seasons, increased extreme weather events, and threats to crop yields.
- Rising temperatures, changing precipitation patterns, and more frequent pests and diseases pose significant challenges to food security and agricultural sustainability worldwide.

8.4.6 Switching Between Different LLMs

These days, many NLP applications require switching amid multiple large language models (LLMs). LLMs are continuously evolving and improving with every passing day—significantly enhancing their ability to handle diverse tasks. There are a variety of LLMs available from multiple companies. Each model comes with its unique strengths and also equally unique limitations. Choosing the right model for the right task is crucial. In this context, the ability to seamlessly switch between LLMs enhances flexibility, efficiency, and overall performance in NLP applications.

Basic architecture and training datasets are the primary considerations when choosing between LLMs. These factors have a profound impact on any LLM's ability to understand language, model response quality, and computational resource requirements. GPT-based models outrival conversational AI, while BERT is more appropriate for document classification tasks. If developers can dynamically switch between these models, they can optimal NLP applications for a variety of tasks. This way, the developer can appropriately leverage the strengths of each model. These hybrid tactics, which combine multiple LLMs, can be an added advance for the accuracy and reliability of complex workflows.

Despite several advantages, switching between LLMs presents multiple technical challenges,, specifically those concerning model compatibility and resource management. Developers need to establish an effective integration strategy that can handle variations in input-output formats, tokenization techniques, and LLM-specific parameters. As the NLP landscape continues to improve over time, learning the art of switching between multiple LLMs will become increasingly indispensable for realizing the full potential of AI-powered language skills.

Currently, LLM switching technologies are in their early stages of development. Still, seamless transitions between LLMs and optimum resource management across multiple models currently remain the areas for improvement. Under these circumstances, next, we provide a basic code demo that demonstrates integration and switching between different LLMs via APIs. Advanced demos that involve seamless model switching and resource optimization usually require custom-built architecture and additional tools.

Switching Between GPT-4 and GPT-3.5 via OpenAI API

The program shown in Listing 8-14 demonstrates how to integrate and switch between two different LLMs—OpenAI's GPT and Hugging Face's BERT—using their respective APIs. It allows users to select a model for a specific task, process the input, and receive the corresponding output.

- This program allows users to choose between OpenAI's GPT for text generation and Hugging Face's BERT for sentiment analysis, enabling them to process input based on their selected model.

- Accepts user input based on the selected task and processes it using the chosen model.

- Outputs the generated text (from GPT) or sentiment analysis result (from BERT).

Note This demo is done in `openai==0.28`. So, to run the code, you need to run

`!pip install openai==0.28`

You can create a separate Python environment and run it there.

Listing 8-14. Switching Between GPT-4 and GPT-3.5 via OpenAI API

```
# !pip install openai==0.28
# Read the open ai API key from your text file
f = open('C:\\Users\\Shailendra Kadre\\Desktop\\OPEN_AI_
KEY.txt')
api_key = f.read()
import openai

# Set your API key
openai.api_key = api_key

# Function to generate a response with model switching
def generate_response(model="gpt-3.5-turbo"):
    try:
        # Use the selected model to generate a response
        response = openai.ChatCompletion.create(
            model=model,  # Model selection here
            messages=[{"role": "user", "content": "Hello, 
            world!"}]
        )
        
        # Extract and print the response
        print(f"Response from {model}: {response['choices'][0]
        ['message']['content']}")
    
    except Exception as e:
        print(f"An error occurred: {e}")

# Switch between different models
models = ["gpt-3.5-turbo", "gpt-4"]
```

```
for model in models:
    print(f"\nSwitching to model: {model}")
    generate_response(model)
```

Switching to model: gpt-3.5-turbo
Response from gpt-3.5-turbo: Hello there! How can I assist you today?

Switching to model: gpt-4
Response from gpt-4: Hello! How can I assist you today?

8.5 Chapter Recap

This chapter examined the transformative world of contemporary NLP, which originated with the surge of transformers. The chapter began with self- and cross-attention mechanisms and the evolution of NLP from traditional methods to revolutionary models like BERT. We discussed DistilBERT and then extended the journey to multimodal NLP. Then, we showcased how OpenAI's CLIP seamlessly integrates text, images, and audio, which allows NLP tasks like text-to-image search with amazing accuracy.

The chapter then delves into the evolution of text generation, from RNNs to GPT, revealing the intelligent processes behind GenAI models. All of it highlights the adaptability of GPT in its real-life applications. Later, we examined the emergent trends in AI-driven creativity. Building on this groundwork, we conducted multiple code demos using LangChain and OpenAI APIs, which are revolutionizing the technology world. Using these powerful tools, we demonstrated dynamic chat assistants and advanced query handling. The chapter later worked with examples of seamless interfacing with LLMs and built innovative applications in text embeddings, vector-based search, and context-aware responses.

CHAPTER 8 TRANSFORMERS, GENERATIVE AI, AND LANGCHAIN

Toward the end of this chapter, learners are also steered through prompt crafting, serialization techniques, and customization of inputs and outputs (to and from LLMs) for custom-made applications. A step-by-step walkthrough of switching between GPT-4 and GPT-3.5 illustrates the flexibility of LLMs. This chapter opens the door to the future of NLP, combining theoretical insights with real-world applications for aspiring AI practitioners.

CHAPTER 9

Advancing with LangChain and OpenAI

The evolution of AI is significantly adding to the ability of business applications to handle complex workflows, retrieve relevant data efficiently, and maintain context across interactions. LangChain's cutting-edge framework delivers critical tools for the integration of external data sources, constructing multi-step AI workflows, and deploying scalable AI-powered solutions. This chapter explores how LangChain empowers seamless data connections, structured processing through chains, effective memory management, interactive agents, and scalable deployment strategies.

The first section looks into **data connection**. It specifies how LangChain relates to both structured and unstructured data through the use of document loaders, transformers, vector databases like ChromaDB, and multimodal sources. Next, we attend to **chains** and describe how to design and develop simple and multi-step AI workflows, handle function calls, and optimize error handling. **Memory management** in LangChain is critical for AI-powered applications, and this chapter explores several memory types, tracking interactions, and improving conversational context. Later, we discuss how AI agents can dynamically select tools, automate tasks, and adapt to complex domains. Toward the end of this chapter, we examine various aspects of **deployment and scalability** for LangChain applications. This section offers insights into deploying LangChain applications in production environments while ensuring system performance and long-term system maintainability.

After completing this chapter, developers should be able to build AI-powered applications that are intelligent, efficient, context-aware, and adaptable to diverse business and other use cases. This chapter focuses chiefly on practical code examples, prevalent industry best practices, and optimization techniques that can streamline development and support the development and deployment of increasingly real-world AI executions.

9.1 Data Connection with Application-Specific Data Sources

In natural language processing (NLP) applications or any application for that matter, effective data connection is critical for retrieving, processing, and using text or other data types from multiple sources. This section explores the vital components of integrating AP-powered applications into specific data sources to ensure seamless document loading, transformation, and embedding for advanced analytics. Let's start with document loaders and integration.

9.1.1 Document Loading, Transformation, and Integration

LangChain document loaders process and extract text (and other information) from a variety of input file formats like CSV, HTML, PDFs, Word docs, and web pages. Using document loaders, we can easily import data into NLP systems or other AI applications. Every loader in LangChain is equipped to handle a specific document type. This ensures efficient extraction and pre-processing of text. For instance, PyMuPDFLoader is specially designed to work with PDF files, while UnstructuredLoader can work with a variety of formats.

These LangChain loaders are used for a variety of NLP tasks, including conversational chatbots, text summarization tools, and search engines. They are designed for converting unstructured text data into a structured format that can be understood and analyzed by NLP models. A couple of loaders also have support for chunking. Document loaders in LangChain significantly save time and effort in NLP developments by simplifying data handling.

Next, we demonstrate three basic document loaders.

LangChain File Loaders for PDF, HTML, and CSV

This code in Listing 9-1 loads and extracts content from PDF, HTML, and CSV files using specific loaders for each file type. It prints the first 500 characters of the content from each document to the console for preview. The process is repeated for all file types: PDF, HTML, and CSV.

Listing 9-1. LangChain File Loaders for PDF, HTML, and CSV

```
# Install these packages if not done already
#!pip install pymupdf
#!pip install Unstructured
from langchain.document_loaders import PyMuPDFLoader  # Import PDF loader
from langchain.document_loaders import UnstructuredHTMLLoader  # Import HTML loader
from langchain.document_loaders import CSVLoader  # Import CSV loader
```

```python
# Paths for sample csv, pdf, and html files
csv_path = "sample_csv.csv"
pdf_path = "sample_pdf.pdf"
html_path = "sample_html.pdf"

def load_and_display_pdf(pdf_path):
    # Initialize the PDF loader with the file path
    loader = PyMuPDFLoader(pdf_path)
    # Load the document into a list of pages
    documents = loader.load()
    # Iterate through each page and print the first 500 characters
    for i, doc in enumerate(documents):
        print(f"--- Page {i + 1} ---\n")
        print(doc.page_content[:500])  # Print extracted text snippet
        print("\n" + "-" * 40 + "\n")  # Separator for readability

def load_and_display_html(html_path):
    # Initialize the HTML loader with the file path
    loader = UnstructuredHTMLLoader(html_path)
    # Load the document
    documents = loader.load()
    # Iterate through each document and print the first 500 characters
    for i, doc in enumerate(documents):
        print(f"--- HTML Document {i + 1} ---\n")
        print(doc.page_content[:500])  # Print extracted text snippet
        print("\n" + "-" * 40 + "\n")  # Separator for readability
```

CHAPTER 9   ADVANCING WITH LANGCHAIN AND OPENAI

```python
def load_and_display_csv(csv_path):
    # Initialize the CSV loader with the file path
    loader = CSVLoader(csv_path)
    # Load the document into rows
    documents = loader.load()
    # Iterate through each row and print the first 500 characters
    for i, doc in enumerate(documents):
        print(f"--- CSV Row {i + 1} ---\n")
        print(doc.page_content[:500])  # Print extracted text snippet
        print("\n" + "-" * 40 + "\n")  # Separator for readability

if __name__ == "__main__":
    # Define file paths
    pdf_path = "sample_pdf.pdf"
    html_path = "sample_html.html"
    csv_path = "sample_csv.csv"

    # Load and display content from each file type
    print("Loading pdf file......\n")
    load_and_display_pdf("sample_pdf.pdf")
    print("Loading html file......\n")
    load_and_display_html(html_path)
    print("Loading csv file......\n")
    load_and_display_csv(csv_path)
```
OUTPUT:
Loading pdf file......

CHAPTER 9   ADVANCING WITH LANGCHAIN AND OPENAI

```
--- Page 1 ---

Lorem ipsum
Lorem ipsum dolor sit amet, consectetur adipiscing
elit. Nunc ac faucibus odio......continued...
----------------------------------------
```

Loading html file......

```
--- HTML Document 1 ---

Sample HTML 1

Minime vero, inquit ille, consentit....continued...

----------------------------------------

--- CSV Row 2 ---

Industry: Advertising/Public Relations
---csv continued.....
```

LangChain is equipped with influential tools for integrating and processing a variety of input documents. These tools enable **seamless data extraction and required data manipulations**. Next, we explore how to use LangChain's document loaders to load and parse various file formats. We can leverage these tools to effortlessly extract appropriate information and prepare the documents for further analysis or machine learning tasks.

Document Transformers: Splitting Text for NLP Processing

- **Overview**: Demonstrates how to split text into sentences, words, and characters using LangChain.

- **File handling**: Loads text from sample_text.txt using LangChain's TextLoader.

CHAPTER 9 ADVANCING WITH LANGCHAIN AND OPENAI

- **Sentence splitting**: Uses CharacterTextSplitter with a period separator.

- **Token splitting**: Splits words based on whitespace.

- **Character splitting**: Converts text into individual characters for fine-grained analysis.

Listing 9-2 below depicts the program for Splitting Text for NLP Processing Using Document Transformers

Listing 9-2. Code Demo for Document Transformers

```
# IMPORT NECESSARY MODULES
from langchain.document_loaders import TextLoader
from langchain.text_splitter import CharacterTextSplitter

# LOAD TEXT FROM FILE USING LANGCHAIN'S DOCUMENT LOADER
loader = TextLoader("sample_text.txt", encoding="utf-8")
documents = loader.load()
document_text = documents[0].page_content  # EXTRACT
TEXT CONTENT

# FUNCTION TO SPLIT TEXT INTO SENTENCES
def split_by_sentence(text):
    text_splitter = CharacterTextSplitter(separator=". ",
    chunk_size=1000)
    return text_splitter.split_text(text)

# FUNCTION TO SPLIT TEXT INTO WORDS
def split_by_word(text):
    return text.split()  # SIMPLE WORD SPLIT USING SPACE

# FUNCTION TO SPLIT TEXT INTO CHARACTERS
```

CHAPTER 9   ADVANCING WITH LANGCHAIN AND OPENAI

```
def split_by_character(text):
    return list(text)  # CONVERT STRING TO LIST OF CHARACTERS

# DEMONSTRATING THE SPLITS
print("\n--- SPLIT BY PERIOD (SENTENCES) ---")
print(split_by_sentence(document_text))

print("\n--- SPLIT BY TOKEN (WORDS) ---")
print(split_by_word(document_text))

print("\n--- SPLIT BY CHARACTER ---")
print(split_by_character(document_text))
```
OUTPUT:

--- SPLIT BY PERIOD (SENTENCES) ---
['LangChain is a framework for developing applications powered by large language models. \nIt helps with data retrieval, memory, and document processing. \nAI agents use LangChain to handle conversations and reasoning.']

--- SPLIT BY TOKEN (WORDS) ---
['LangChain', 'is', 'a', 'framework', 'for', 'developing', 'applications', 'powered', 'by', 'large', 'language', 'models.', 'It', 'helps', 'with', 'data', 'retrieval,', 'memory,', 'and', 'document', 'processing.', 'AI', 'agents', 'use', 'LangChain', 'to', 'handle', 'conversations', 'and', 'reasoning.']

--- SPLIT BY CHARACTER ---
['L', 'a', 'n', 'g', 'C', 'h', 'a', 'i', 'n', ' ', 'i', 's', ' ', 'a', ' ', 'f', 'r', 'a', 'm', 'e', 'w', 'o', 'r', 'k', ' ', 'f', 'o', 'r', ' ', 'd', 'e', 'v', 'e', 'l', 'o', 'p', 'i', 'n', 'g', ' ', 'a', 'p', 'p', 'l', 'i', 'c', 'a', 't', 'i', 'o', 'n', 's', ' ', 'p', 'o', 'w', 'e', 'r', 'e', 'd', ' '

CHAPTER 9 ADVANCING WITH LANGCHAIN AND OPENAI

', 'b', 'y', ' ', 'l', 'a', 'r', 'g', 'e', ' ', 'l', 'a', 'n', 'g', 'u', 'a', 'g', 'e', ' ', 'm', 'o', 'd', 'e', 'l', 's', '.', ' ', ' ', '\n', 'I', 't', ' ', 'h', 'e', 'l', 'p', 's', ' ', 'w', 'i', 't', 'h', ' ', 'd', 'a', 't', 'a', ' ', 'r', 'e', 't', 'r', 'i', 'e', 'v', 'a', 'l', ',', ' ', 'm', 'e', 'm', 'o', 'r', 'y', ',', ' ', 'a', 'n', 'd', ' ', 'd', 'o', 'c', 'u', 'm', 'e', 'n', 't', ' ', 'p', 'r', 'o', 'c', 'e', 's', 's', 'i', 'n', 'g', '.', ' ', ' ', '\n', 'A', 'I', ' ', 'a', 'g', 'e', 'n', 't', 's', ' ', 'u', 's', 'e', ' ', 'L', 'a', 'n', 'g', 'C', 'h', 'a', 'i', 'n', ' ', 't', 'o', ' ', 'h', 'a', 'n', 'd', 'l', 'e', ' ', 'c', 'o', 'n', 'v', 'e', 'r', 's', 'a', 't', 'i', 'o', 'n', 's', ' ', 'a', 'n', 'd', ' ', 'r', 'e', 'a', 's', 'o', 'n', 'i', 'n', 'g', '.', '\n', '\n', '\n', '\n']

9.1.2 Chains: Construct Sequences of Calls for Specific Tasks

The LLM Chains framework offers automation and optimization of workflows by connecting several AI-driven steps to perform complex tasks. In simple words, LLM Chains are used to connect different steps when using LLMs. LLM Chains help developers by breaking down big and complex tasks into more maintainable, smaller parts; this makes AI more accurate and useful in business applications, such as customer-facing chatbots, search engines, and work automation tools. LangChain helps developers integrate LLMs competently with any structured logic—be it technical or business applications. Modular applications, as in LLM Chains, enable better control and customization in applications by streamlining workflows.

AI application developers utilize LLM Chains to empower stepwise reasoning, prompt engineering, and contextual memory across

CHAPTER 9   ADVANCING WITH LANGCHAIN AND OPENAI

interactions, thereby creating more reliable and adaptive AI solutions. Businesses leverage LLM Chains for developing personalized recommendations, predictive analytics, and fraud detection. With its fast-growing adoption, LLM Chains is poised to transform multiple domains, including finance, healthcare, and e-commerce, with scalable AI-driven solutions. Figure 9-1 is an easy-to-understand schematic of how LLM Chains work in the context of business and scientific applications.

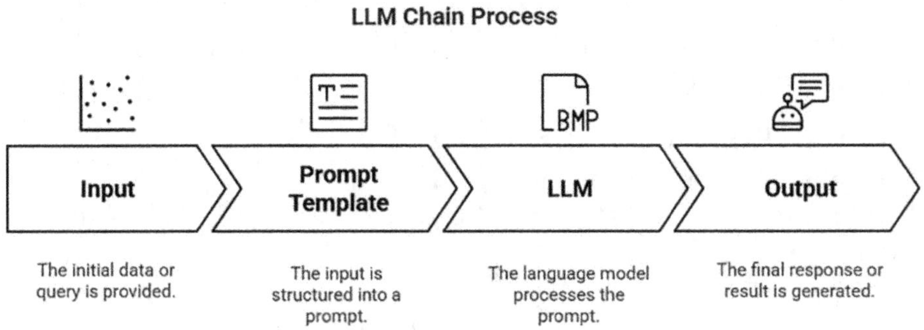

LLM Chain Process

Input	Prompt Template	LLM	Output
The initial data or query is provided.	The input is structured into a prompt.	The language model processes the prompt.	The final response or result is generated.

Figure 9-1. *Schematic of LLM Chains*

Next, we examine code demos that use LLM Chains for the development of useful applications for businesses.

Business Document Processing: Summarization, Keyword Extraction, and Sentiment Analysis

This section focuses on processing different types of business documents to help understand their content quickly. It uses advanced tools to summarize important information from files like PDFs and spreadsheets. It also finds key words that highlight important topics for business decisions. Additionally, it analyzes customer feedback to understand their feelings and satisfaction levels. These automated methods make handling large amounts of business data faster and more accurate. The code demo of Business Document Processing is given in Listing 9-3.

CHAPTER 9   ADVANCING WITH LANGCHAIN AND OPENAI

- Loads and processes diverse business documents (PDF, HTML, CSV) for analysis.

- Summarizes document content for quick business insights using GPT-3.5.

- Extracts key business-relevant keywords to aid decision-making.

- Performs sentiment analysis on customer feedback to gauge satisfaction.

- Provides efficient business document analysis through automated text processing.

Listing 9-3. Program to Demonstrate Business Document Processing

```
# Read the open ai api key from your text file
f = open('C:\\Users\\Shailendra Kadre\\Desktop\\OPEN_AI_KEY.txt')
api_key = f.read()
import os
from langchain.document_loaders import PyPDFLoader,
UnstructuredHTMLLoader, CSVLoader
from langchain.text_splitter import CharacterTextSplitter
from langchain_openai import ChatOpenAI   # Correct import for
OpenAI Chat model
from langchain.prompts import PromptTemplate
from langchain.chains import LLMChain

# Load and parse documents
pdf_loader = PyPDFLoader("sample_pdf.pdf")
pdf_docs = pdf_loader.load()

html_loader = UnstructuredHTMLLoader("sample_html.html")
html_docs = html_loader.load()
```

```
csv_loader = CSVLoader("sample_csv.csv")
csv_docs = csv_loader.load()

# Text Splitting for Processing
splitter = CharacterTextSplitter(chunk_size=500, chunk_
overlap=50)
pdf_chunks = splitter.split_documents(pdf_docs)
html_chunks = splitter.split_documents(html_docs)
csv_chunks = splitter.split_documents(csv_docs)

# Initialize the ChatOpenAI model with API key
llm = ChatOpenAI(model="gpt-3.5-turbo", openai_api_key=api_key)

# Summarization using OpenAI (Using LLMChain directly)
summary_prompt = PromptTemplate(
    input_variables=["text"],
    template="Summarize the following business
    document:\n{text}"
)
summary_chain = LLMChain(llm=llm, prompt=summary_prompt)

# Using .invoke() instead of .run()
pdf_summary = summary_chain.invoke({"text": pdf_chunks[0].page_
content})
html_summary = summary_chain.invoke({"text": html_chunks[0].
page_content})
csv_summary = summary_chain.invoke({"text": csv_chunks[0].page_
content})

print("\n--- PDF Summary ---\n", pdf_summary)
print("\n--- HTML Summary ---\n", html_summary)
print("\n--- CSV Summary ---\n", csv_summary)
```

CHAPTER 9 ADVANCING WITH LANGCHAIN AND OPENAI

```python
# Keyword Extraction using OpenAI (Using LLMChain directly)
keyword_prompt = PromptTemplate(
    input_variables=["text"],
    template="Extract the top 5 keywords from the following
    document:\n{text}"
)
keyword_chain = LLMChain(llm=llm, prompt=keyword_prompt)

# Using .invoke() instead of .run()
pdf_keywords = keyword_chain.invoke({"text": pdf_chunks[0].
page_content})
html_keywords = keyword_chain.invoke({"text": html_chunks[0].
page_content})
csv_keywords = keyword_chain.invoke({"text": csv_chunks[0].
page_content})

print("\n--- PDF Keywords ---\n", pdf_keywords)
print("\n--- HTML Keywords ---\n", html_keywords)
print("\n--- CSV Keywords ---\n", csv_keywords)

# Sentiment Analysis for Business Insights (Using LLMChain
directly)
sentiment_prompt = PromptTemplate(
    input_variables=["text"],
    template="Analyze the sentiment of the following
    customer feedback and rate as Positive, Neutral, or
    Negative:\n{text}"
)
sentiment_chain = LLMChain(llm=llm, prompt=sentiment_prompt)

csv_sentiment = sentiment_chain.invoke({"text": csv_chunks[0].
page_content})
```

CHAPTER 9   ADVANCING WITH LANGCHAIN AND OPENAI

```
# Print the sentiment analysis output
print("\n--- CSV Sentiment Analysis ---\n", csv_sentiment['text'])
```

--- PDF Summary ---
 {'text': 'The document discusses Lorem ipsum dolor sit amet, consectetur adipiscing elit. It mentions various business activities such as marketing, sales, and customer service. It also includes a table with rows and columns for data representation.'}

--- HTML Summary ---
 {'text': 'The document discusses various topics, such as comparing medicine and governance to wisdom, the importance of speaking in a customary manner when discussing something, and whether enduring suffering increases happiness. It also raises questions about whether jokes, secrets, and hidden truths should be shared with everyone. The document concludes with a question about whether prolonged suffering ultimately leads to greater happiness.'}

--- CSV Summary ---
 {'text': 'The document likely discusses information related to the accounting and finance industry, which may include financial reporting, taxation, auditing, and other financial services. It may also cover industry trends, regulatory updates, and best practices for financial management.'}

--- PDF Keywords ---
 {'text': '1. Maecenas\n2. condimentum\n3. ipsum\n4. fringilla\n5. ligula'}

--- HTML Keywords ---
 {'text': '1. Lorem\n2. Ipsum\n3. Dolor\n4. Sit\n5. Amet'}

--- CSV Keywords ---
{'text': '1. Accounting\n2. Finance\n3. Industry\n4. Top\n5. Keywords'}

--- CSV Sentiment Analysis ---
"Great experience working with this accounting firm. Their team was knowledgeable, efficient, and professional. Will definitely be using their services again in the future."

Sentiment: Positive

9.1.3 Text Embeddings and Vector Databases

The NLP eco space has been transformed since the advent of text embeddings. Embeddings have changed the way machines interpret and process language. Conversion of text into high-dimensional numeric vectors helps to capture semantic relationships. This facilitates sophisticated text comprehension. NLP applications can't capture contextual meanings using traditional keyword-based searches. But embeddings have successfully filled this gap. Text embeddings allow NLP models to retrieve information based on intent rather than exact wording. Understanding context becomes indispensable when it comes to tasks like semantic search, document similarity analysis, and intelligent retrieval systems.

Vector databases like ChromaDB can efficiently store and manage these text embeddings with adequate scaling capabilities. Vector databases help us in enabling quick and accurate similarity searches. This makes them best suited for retrieval-augmented generation (RAG) systems that are useful in AI-powered assistants and knowledge management systems. By integrating LangChain with OpenAI models, developers can effortlessly generate, store, and query embeddings for advanced NLP tasks. AI-driven applications are still evolving. Text embeddings and vector databases are going to have a central role in progressing smart search, recommender systems, and conversational AI.

CHAPTER 9   ADVANCING WITH LANGCHAIN AND OPENAI

Text Similarity Using OpenAI Embeddings and Cosine Similarity

Listing 9-4 below dipicts the program for Calculating Text Similarity Using OpenAI Embeddings and Cosine Similarity

Listing 9-4. Text Similarity Using OpenAI Embeddings and Cosine Similarity

```
# A simple demo of **text embeddings and similarity search
using `OpenAI` embeddings **without using ChromaDB.
# This will show how text embeddings work without requiring a
vector database.
"""

- Program: Text Embeddings & Similarity Search (No ChromaDB)
- Steps:
1. Convert text into **embeddings** using `OpenAI` API.
2. Store embeddings in a list (instead of ChromaDB).
3. Use **Cosine Similarity** to find similar texts.
"""
'\n\n- Program: Text Embeddings & Similarity Search (No
ChromaDB)\n- Steps:    \n1. Convert text into **embeddings**
using `OpenAI` API.   \n2. Store embeddings in a list (instead
of ChromaDB).   \n3. Use **Cosine Similarity** to find similar
texts.\n\n'
# Read the open ai API key from your text file
f = open('C:\\Users\\Shailendra Kadre\\Desktop\\OPEN_AI_KEY.txt')
api_key = f.read()
# Importing necessary libraries
import openai   # OpenAI's Python client library to interact
with its API
```

CHAPTER 9 ADVANCING WITH LANGCHAIN AND OPENAI

```python
import numpy as np  # NumPy for handling numerical computations and arrays
from sklearn.metrics.pairwise import cosine_similarity
# Function to compute similarity between vectors

# Set your OpenAI API key (Replace 'api_key' with your actual API key)
openai.api_key = api_key  # This allows access to OpenAI's services

# Function to generate text embeddings using OpenAI's embedding model
def get_embedding(text):
    response = openai.embeddings.create(  # Request embeddings from OpenAI
        model="text-embedding-ada-002",  # Specify the embedding model
        input=[text]  # Input text must be passed as a list
    )
    return np.array(response.data[0].embedding)  # Extract and return the embedding as a NumPy array

# List of sample text data for embedding generation
texts = [
    "Artificial Intelligence is transforming industries.",
    # AI-related sentence
    "Machine learning helps in predictive analytics.",
    # ML-related sentence
    "Deep learning is a subset of machine learning.",
    # DL-related sentence
    "I love pizza and Italian food."  # Unrelated topic (food preference)
]
```

CHAPTER 9 ADVANCING WITH LANGCHAIN AND OPENAI

```
# Convert each text in the list into an embedding using the
get_embedding function
embeddings = [get_embedding(text) for text in texts]

# Convert the list of embeddings into a NumPy array for
efficient processing
embeddings_matrix = np.array(embeddings)

# Compute cosine similarity between the first text embedding
and all other embeddings
similarities = cosine_similarity([embeddings_matrix[0]],
embeddings_matrix)

# Print similarity scores for each text compared to the
first text
print("Similarity Scores with First Text:")
for i, score in enumerate(similarities[0]):  # Iterate over
similarity scores
    print(f"{i}: {score:.4f} → {texts[i]}")  # Print index,
similarity score, and corresponding text
Similarity Scores with First Text:
0: 1.0000 → Artificial Intelligence is transforming
industries.
1: 0.8563 → Machine learning helps in predictive analytics.
2: 0.8294 → Deep learning is a subset of machine learning.
3: 0.7261 → I love pizza and Italian food.
"""
```

- How It Works:
1. Gets embeddings for each text using OpenAI's `text-embedding-ada-002`.
2. Stores embeddings in a list (instead of ChromaDB).
3. Computes similarity using `cosine_similarity()` from `sklearn`.

CHAPTER 9 ADVANCING WITH LANGCHAIN AND OPENAI

4. *Prints similarity scores, showing which texts are most similar.*

"""

'\n\n- How It Works:\n1. Gets embeddings for each text using OpenAI's `text-embedding-ada-002`. \n2. Stores embeddings in a list (instead of ChromaDB). \n3. Computes similarity using `cosine_similarity()` from `sklearn`. \n4. Prints similarity scores, showing which texts are most similar. \n\n'

"""

- *Expected Output:*
Similarity Scores with First Text:
0: 1.0000 → Artificial Intelligence is transforming industries.
1: 0.8643 → Machine learning helps in predictive analytics.
2: 0.7892 → Deep learning is a subset of machine learning.
3: 0.1125 → I love pizza and Italian food.
```

- *Higher scores mean more similarity. Unrelated text (pizza) gets a low score.*

"""

'\n- Expected Output:\nSimilarity Scores with First Text:\n0: 1.0000 → Artificial Intelligence is transforming industries.\ n1: 0.8643 → Machine learning helps in predictive analytics.\ n2: 0.7892 → Deep learning is a subset of machine learning.\ n3: 0.1125 → I love pizza and Italian food.\n```\n- Higher scores mean more similarity. Unrelated text (pizza) gets a low score.\n\n'
```

CHAPTER 9   ADVANCING WITH LANGCHAIN AND OPENAI

9.2 AI Agents

LLM-based AI agents are intelligent systems powered by LLMs that can understand, generate, and process human language. They automate tasks, provide insights, and interact with users naturally. Such AI agents can be used to develop advanced chatbots, code assistants, automated content generators, customer support, data analysis, personalized recommendations, and process automation. These autonomous AI agents improve efficiency, reduce costs, and enable smarter decision-making. Next, we examine a code demo to showcase the working of these agents. This code demo is self-sufficient in terms of explaining the code in sufficient detail.

LangChain-Based Stock Analysis Agent

- **IMPORT MODULES** loads required modules for environment setup, data handling, and AI integration.

- **SET API KEY** stores the OpenAI API key in an environment variable.

- **DEFINE STOCK LIST** contains 50 real-world stock symbols.

- **GENERATE STOCK DATA** creates randomized stock market data for 10 days.

- **CREATE DATAFRAME** stores stock data in a pandas dataframe.

- **FETCH STOCK DATA** is a function that returns historical data for a given stock.

- **ANALYZE STOCK TREND** computes a simple moving average (SMA) and generates a buy/sell/hold signal.

CHAPTER 9   ADVANCING WITH LANGCHAIN AND OPENAI

- **DEFINE LANGCHAIN TOOLS**
  - tool to fetch stock data
  - tool to analyze stock trends
- **INITIALIZE OPENAI LLM** creates an AI model instance with a set temperature.
- **CREATE AI AGENT** integrates LLM with the defined tools using LangChain.
- **EXECUTE QUERY** runs an agent command to analyze stock trends for a specific symbol (e.g., AAPL).

```
# Read the open ai API key from your text file
f = open('C:\\Users\\Shailendra Kadre\\Desktop\\OPEN_AI_KEY.txt')
api_key = f.read()
import os  # Import the os module to set environment variables
import random  # Import random module for generating dummy stock prices
from datetime import datetime, timedelta  # Import datetime for handling dates
from langchain.llms import OpenAI  # Import OpenAI LLM from LangChain
from langchain.agents import initialize_agent, AgentType
# Import agent tools from LangChain
from langchain.tools import Tool  # Import Tool class to create custom LangChain tools
import pandas as pd  # Import pandas for data handling and analysis

# Set up your OpenAI API key (replace with your actual key)
os.environ["OPENAI_API_KEY"] = api_key
```

```python
# List of 50 real-world stock symbols
stocks = ["AAPL", "GOOGL", "AMZN", "TSLA", "MSFT", "NFLX",
"META", "NVDA", "BABA", "DIS", "V", "JPM", "PYPL", "MA", "INTC",
"IBM", "ORCL", "CSCO", "ADBE", "AMD", "UBER", "LYFT", "SQ",
"SHOP", "TWTR", "SNAP", "PINS", "ZM", "DOCU", "ROKU", "BA",
"GE", "CAT", "MMM", "F", "GM", "NKE", "KO", "PEP", "MCD","WMT",
"TGT", "HD", "LOW", "COST", "PG", "JNJ", "MRNA", "PFE", "BMY"
]

# Generate dummy stock market data
num_days = 10  # Number of days per stock
start_date = datetime(2024, 1, 1)  # Start date
data = []  # Initialize empty list to store stock data

# Generate data for each stock over 10 days
for stock in stocks:
    for i in range(num_days):
        date = start_date + timedelta(days=i)  # Increment date
        open_price = round(random.uniform(100, 1000), 2)
        # Random open price
        high_price = round(open_price + random.uniform(5, 50),
        2)  # High slightly above open
        low_price = round(open_price - random.uniform(5, 50),
        2)  # Low slightly below open
        close_price = round(random.uniform(low_price, high_
        price), 2)  # Close within range
        volume = random.randint(500000, 50000000)  # Random
        trading volume

        # This line adds a new row to a list called data
        data.append([date.strftime("%Y-%m-%d"), stock, open_
        price, high_price, low_price, close_price, volume])
        # date.strftime("%Y-%m-%d") - Formats the date object
        into a YYYY-MM-DD string.
```

```python
# Create DataFrame
df = pd.DataFrame(data, columns=["Date", "Stock", "Open",
"High", "Low", "Close", "Volume"])

# Function to get stock data for a specific symbol
def get_stock_data(symbol: str):
    return df[df["Stock"] == symbol].reset_index(drop=True)
    # Filter DataFrame for the given stock symbol

# Function to analyze stock trends based on SMA (Simple Moving Average)
def analyze_trend(stock_data):
    stock_data["SMA"] = stock_data["Close"].rolling(window=3).mean()  # Calculate Simple Moving Average (SMA)
    latest_price = stock_data["Close"].iloc[-1]  # Get the latest closing price
    latest_sma = stock_data["SMA"].iloc[-1]  # Get the latest SMA value
    """
    `latest_price = stock_data["Close"].iloc[-1]`

    ### Explanation:
    - `stock_data["Close"]`: Extracts the "Close" column from the `stock_data` DataFrame, which contains closing prices of the stock.
    - `.iloc[-1]`: Selects the last row of the "Close" column, retrieving the most recent closing price.

    ### Purpose:
    This line gets the latest closing price of a stock from the given dataset.
    """
```

```python
    # Generate trading signal based on SMA comparison
    if latest_price > latest_sma:
        return "Buy signal: Price is above SMA"
    elif latest_price < latest_sma:
        return "Sell signal: Price is below SMA"
    else:
        return "Hold signal: Price is at SMA"

# Define a LangChain tool to fetch stock data
stock_data_tool = Tool(
    name="Stock Data Fetcher",  # Name of the tool
    func=lambda symbol: get_stock_data(symbol).to_string(),
    # Function to execute
    description="Fetches dummy stock data for a given
    symbol"  # Description of the tool
)

# Define a LangChain tool to analyze the stock trend
trend_analysis_tool = Tool(
    name="Stock Trend Analyzer",  # Name of the tool
    func=lambda symbol: analyze_trend(get_stock_
    data(symbol)),  # Function to execute
    description="Analyzes stock trend and gives buy/sell
    signals"  # Description of the tool
)

# Initialize the LangChain LLM
llm = OpenAI(api_key=os.environ["OPENAI_API_KEY"],
temperature=0.7)  # Create an instance of OpenAI model with
moderate randomness

# Create the AI agent using LangChain
agent = initialize_agent(
```

```
    tools=[stock_data_tool, trend_analysis_tool],
    # List of tools the agent can use
    llm=llm,  # Assign the LLM to the agent
    agent=AgentType.ZERO_SHOT_REACT_DESCRIPTION,
    # Specify the type of agent
    verbose=True  # Enable verbose output for debugging
)

# Example usage (replace with actual stock symbol)
response = agent.run("Analyze the trend for AAPL")
# Run the agent with a query
print(response)  # Print the agent's response
> Entering new AgentExecutor chain...
 I should use the Stock Data Fetcher to get the stock data
for AAPL
Action: Stock Data Fetcher
Action Input: AAPL
Observation:     Date Stock  Open    High    Low    Close    Volume
0  2024-01-01  AAPL  950.12  957.85  900.64  901.40  15083224
1  2024-01-02  AAPL  525.51  549.66  477.65  518.36  14434448
2  2024-01-03  AAPL  907.62  919.88  863.87  898.99  15286254
3  2024-01-04  AAPL  857.77  875.54  849.98  857.14  35041104
4  2024-01-05  AAPL  941.17  974.03  924.63  934.36  33134796
5  2024-01-06  AAPL  552.19  565.87  529.39  546.66   2792180
6  2024-01-07  AAPL  953.96  960.45  943.50  958.87  41613530
7  2024-01-08  AAPL  482.18  525.64  469.70  485.87  14577853
8  2024-01-09  AAPL  341.44  369.70  319.75  367.96  23713339
9  2024-01-10  AAPL  546.23  564.76  536.35  550.44  27605642
```

Thought: Now that I have the stock data, I can use the Stock Trend Analyzer to analyze the trend for AAPL
Action: Stock Trend Analyzer
Action Input: AAPL

CHAPTER 9   ADVANCING WITH LANGCHAIN AND OPENAI

Observation: Buy signal: Price is above SMA
Thought: Based on the buy signal, it seems like a good time to invest in AAPL

Final Answer: It may be a good idea to buy AAPL based on the buy signal from the Stock Trend Analyzer. However, further research and analysis should be done before making any investment decisions.

> Finished chain.
It may be a good idea to buy AAPL based on the buy signal from the Stock Trend Analyzer. However, further research and analysis should be done before making any investment decisions.

The Role of LLM and Agents in This Code

Role of LLM (Large Language Model)

- `llm = OpenAI(api_key=os.environ["OPENAI_API_KEY"], temperature=0.7)`:
  - This initializes the OpenAI language model (LLM) using LangChain.
  - The model is used to process user queries, interpret them, and generate responses.
  - The `temperature=0.7` allows for a balanced mix of deterministic and creative responses.

Role of Agents

- `agent = initialize_agent(...)`:
  - The agent acts as an AI-powered assistant that interacts with users and selects the right tools to answer their queries.

CHAPTER 9 ADVANCING WITH LANGCHAIN AND OPENAI

- It uses ZERO_SHOT_REACT_DESCRIPTION, meaning it decides dynamically which tools to use based on the query.

- The agent can

    a. **Fetch stock data** using stock_data_tool.

    b. **Analyze stock trends** using trend_analysis_tool.

Final Workflow

- When a user asks, **"Analyze the trend for AAPL"**, the agent

    a. **Identifies** that stock analysis is needed.

    b. **Fetches AAPL stock data** using stock_data_tool.

    c. **Runs trend analysis** using trend_analysis_tool.

    d. **Generates a response** (e.g., "Buy signal: Price is above SMA").

9.3 Chapter Recap

This chapter explains the advanced techniques for creating various NLP applications using LangChain and OpenAI. At the beginning of this chapter, the significance of connecting to various data sources is discussed. A detailed explanation is included for integrating document loaders, transforming documents, and working with text embeddings and vector databases like ChromaDB. For building NLP applications, data is available in multimodal data sources. This can include both text and images. Efficient handling of these multimodal data sources is required for optimal data retrieval from multiple data sources. This is achieved by preprocessing and indexing for more efficient querying.

CHAPTER 9   ADVANCING WITH LANGCHAIN AND OPENAI

Next, this chapter introduces the concept of *chains*, which are the sequence of tasks that are designed to achieve specific goals. In this section, the creation of different types of chains, simple and multi-step chains are explained. LangChain is used to handle function calls. Various approaches are also discussed to optimize chains and to handle errors. For showcasing automatic workflows, practical examples are also provided.

In the later part of this chapter, the concept of memory and its management is discussed to allow the applications to retain their state between different runs. Various types of memory are also discussed, along with the management of user and AI interactions. Implementing both short-term and long-term memory solutions improves conversational applications by enhancing contextual understanding.

At the end of the chapter, a separate focus is provided on agents—dynamic tools capable of performing tasks automatically. A detailed explanation is provided for agents on how interactive AI agents work and methods for building domain-specific agents to tackle complex tasks. Detailed guidelines are provided on deploying LangChain applications in production. Their key aspects related to scaling them for high demand and ensuring their reliability through monitoring and maintenance are also covered in this chapter.

CHAPTER 10

Case Study on Symantec Analysis

10.1 Introduction

Since 2006, the world has witnessed a data explosion, which finds an explanation in the "humanity and technology on the second half of the chessboard" (Brynjolfsson and McAfee 2011).[1] Data proliferation has led to an increasing shift in marketing communications, as witnessed on online platforms (Sheth and Kellstadt 2020),[2] along with advocacy by the user community in the form of reviews. This has led to an explosion of unstructured data, primarily in the form of text messages, resulting in inferential statistics being increasingly supplemented by non-inferential approaches, such as natural language processing (NLP), a technique used by computers to understand and generate human language so that large amounts of text data can be analyzed for insights.

This chapter, however, examines semantic analysis—a crucial component of NLP, which helps humans understand the intended meaning of words, moving beyond a mere analysis of a sentence's composition and how companies are leveraging this. Semantic analysis helps in the understanding of the text's main idea, highlighting the relationships between the words used and the sentences constructed.

© Shailendra Kadre, Shailesh Kadre and Subhendu Dey 2025
S. Kadre et al., *Mastering Text Analytics*, https://doi.org/10.1007/979-8-8688-1582-9_10

CHAPTER 10   CASE STUDY ON SYMANTEC ANALYSIS

In other words, it helps humans comprehend written text beyond its grammatical structure by examining context and word relationships to derive deeper insights from the written language.

Finding the intended meaning of words, phrases, and sentences beyond their obvious meaning is nothing new. Semantics finds its roots in the thinking of Plato. In his dialogue Cratylus,[3] Plato talks about words as "symbols of ideas" associated with them instead of being "crude imitations of other natural sounds." He discussed the idea that words have a natural connection to the things they signify. The formal study of semantics began to take shape in the late 19th and early 20th centuries. Michel Bréal, a French philologist, is credited with coining the term "semantics" in 1883.[4] He explored the structure of languages, their evolution over time, and the relationships between them. In simple terms, semantics focuses on the study of language and its meaning. At the same time, semantic analysis enables computers to accurately determine the context of words or phrases that can be interpreted in multiple ways.

In the mid-20th century, the formalization of the study of semantics began to take shape in the field of linguistics. Noam Chomsky's work on generative grammar in the 1950s laid the groundwork for understanding how syntax (the structure of sentences) and semantics (the meaning of sentences) are related. Around the same time, Richard Montague[5] applied mathematical logic to the study of natural language semantics. He proposed Montague semantics, a theory based on natural language semantics and its relation to syntax.

The advent of computers and the development of programming languages in the middle of the 20th century opened new possibilities for the study of semantics. Robert W. Floyd's 1967 paper[6] on programming language semantics is often given credit for introducing the field of programming language semantics. Floyd explained programming languages as having two parts: semantics (i.e., meaning and syntax), and emphasized the importance of precisely encoding these elements so that computers could process them automatically.

It was not until the late 1990s that the Semantic Web concept was proposed by Tim Berners-Lee. This was a major advancement in the field of semantic analysis. The Semantic Web was the first attempt to make web content accessible to machines through structured metadata descriptions and ontologies to give meaning to data. This formed the basis for the development of standards such as the Resource Description Framework (RDF) and the Web Ontology Language (OWL), which are used to represent and reason about data on the web.

It is important to mention that owing to the explosion of text messages, analyzing them to derive meaningful conclusions is not that easy. This is because data proliferation has occurred on a scale where data is in search of techniques, rather than techniques in search of data (Sheth and Kellstadt, 2020).

This is despite advances made in natural language processing in recent years, which have boosted the field of semantic analysis, enabling computers to understand and generate human language, and making it possible to analyze large volumes of text data for valuable insights. Machine learning algorithms, especially deep learning models such as transformers, have significantly improved the accuracy and efficiency of semantic analysis.

Despite techniques trying to catch up with data, there are successful models of companies that use semantic analysis to make insightful decisions. This chapter attempts to explore the future of semantic analysis, which is likely to be more robust and scalable, shaped by advancements in AI and machine learning.

10.2 Methods of Semantic Analysis

Semantic analysis allows computers to understand the right meaning of words or phrases with more than one meaning, which is crucial to the accuracy of text-based NLP applications. The technology not only processes information but also recognizes the connections between pieces of information to decipher massive amounts of data and implement them in the real world. Semantic analysis uses a variety of methods to interpret the meaning of text data. Among the most prominent ones is latent semantic analysis.

10.2.1 Latent Semantic Analysis

Latent semantic analysis (LSA) is a natural language processing method that analyzes relationships between a set of documents and the words within them. LSA is particularly useful for applications like information retrieval and text summarization. It identifies hidden or "latent" associations between terms and concepts in unstructured data. LSA operates on the principle that words appearing in similar contexts tend to share meanings. This allows the technique to infer word significance within text by analyzing patterns of co-occurrence.

LSA is also referred to as latent semantic indexing (LSI). LSA/LSI use singular value decomposition (SVD) to reduce the dimension of the Term-Document matrix, which helps in extracting semantic structures from large document collections.

10.2.1.1 How LSA Works

LSA follows several key steps.

Term-Document Matrix

The process begins with constructing a term-document matrix that captures word frequency across multiple documents. Each row represents a unique word, while each column corresponds to a document. The matrix values indicate the number of times a word appears in a given document.

For example, term "target" appears two times in document X, then the matrix [target, X] = 2. Consider the following three text excerpts to understand the term-document matrix functions.

The salesman achieved the target

The company motivated the salesman

The target achieved helped the company

Document matrix for these sentences may look like Table 10-1.

Table 10-1. Term-Document Matrix

	d1	d2	d3
	1	1	0
...ed	1	0	1
...arget	1	0	1
company	0	1	1
motivated	0	1	0

Singular Value Decomposition

Singular value decomposition (SVD) is a mathematical method used to reduce the dimensionality of the term-document matrix while preserving relationships between words and documents. This transformation condenses the original matrix into a smaller representation that captures essential semantic patterns. SVD is used in LSA to identify hidden relationships between terms and concepts within unstructured data. The following illustration of the use of SVD in Semantic Analysis, with an example of an online retailer analyzing customer reviews, helps yo understand the concept of SVD better.

CHAPTER 10 CASE STUDY ON SYMANTEC ANALYSIS

An online retailer, eMart, hired an analyst to interpret c feedback on its products. They collected customer reviews fers popular products: smart TVs, earbuds, and a smartwatch. The used SVD to identify the key aspects (topics) that customers dis relation to these products.

Step 1. Data Collection and Preprocessing

eMart collects the following simplified customer reviews.

- **Review 1 (smart TV):** "The **picture quality** is **amazing**, and the **sound** is **great**. Very happy with my new **TV!**"

- **Review 2 (smart TV):** "I love the **big screen** and **clear picture**. The **smart features** are very useful."

- **Review 3 (earbuds):** "The **sound quality** is fantastic. These are very **comfortable** to wear."

- **Review 4 (earbuds):** "Great **bass** and noise-canceling. The **battery life** is also excellent in these Earbuds."

- **Review 5 (smartwatch):** "The **fitness tracking** is accurate. The **battery life** is good, and the **watch** looks stylish."

- **Review 6 (smartwatch):** "I use the **heart rate monitor** every day. It's a great **fitness tracker** and a beautiful **watch**."

processing (tokenization, lowercasing, stemming, and words), the following are key terms.

nd	TV	big screen	clear picture
uality	comfortable	bass	noise-canceling
king	watch	heart rate monitor	fitness tracker

Step 2. Creating the Term-Document Matrix

A term-document matrix where rows represent terms and columns represent reviews is shown in Table 10-2.

Table 10-2. Term-Document Matrix

Term	Review 1	Review 2	Review 3	Review 4	Review 5	Review 6
picture quality	1	0	0	0	0	0
sound	1	0	0	0	0	0
tv	1	0	0	0	0	0
big screen	0	1	0	0	0	0
clear picture	0	1	0	0	0	0
smart features	0	1	0	0	0	0
sound quality	0	0	1	0	0	0
comfortable	0	0	1	0	0	0
bass	0	0	0	1	0	0
noise-canceling	0	0	0	1	0	0
battery life	0	0	0	1	1	0
fitness tracking	0	0	0	0	1	1
watch	0	0	0	0	1	1
heart rate monitor	0	0	0	0	0	1
fitness tracker	0	0	0	0	0	1

CHAPTER 10 CASE STUDY ON SYMANTEC ANALYSIS

Step 3. Applying SVD

We apply SVD to the term-document matrix (A).

$A = U\Sigma V^T$

Using Python and NumPy (or similar tools), we perform SVD.

```python
import numpy as np

A = np.array([
    [1, 0, 0, 0, 0, 0],
    [1, 0, 0, 0, 0, 0],
    [1, 0, 0, 0, 0, 0],
    [0, 1, 0, 0, 0, 0],
    [0, 1, 0, 0, 0, 0],
    [0, 1, 0, 0, 0, 0],
    [0, 0, 1, 0, 0, 0],
    [0, 0, 1, 1, 0, 0],
    [0, 0, 0, 1, 0, 0],
    [0, 0, 0, 1, 0, 0],
    [0, 0, 0, 1, 1, 0],
    [0, 0, 0, 0, 1, 1],
    [0, 0, 0, 0, 1, 1],
    [0, 0, 0, 0, 0, 1],
    [0, 0, 0, 0, 0, 1]
])

U, S, VT = np.linalg.svd(A)

print("U Matrix:\n", U)
print("Singular Values:\n", S)
print("VT Matrix:\n", VT)
```

CHAPTER 10   CASE STUDY ON SYMANTEC ANALYSIS

Step 4. Dimensionality Reduction

We decide to keep the top two singular values (k=2). The singular values in S tell us the importance of each dimension (topic). Let's assume after running SVD, the first two singular values are significantly larger than the rest, indicating they capture most of the variance in the data.

We reduce Σ to a k×k matrix. We also truncate U and VT to keep only the first k columns and rows, respectively.

```python
k = 2  # Number of components to keep
U_reduced = U[:, :k]
S_reduced = np.diag(S[:k])
VT_reduced = VT[:k, :]

print("Reduced U Matrix:\n", U_reduced)
print("Reduced Singular Values Matrix:\n", S_reduced)
print("Reduced VT Matrix:\n", VT_reduced)
```

Step 5. Topic Interpretation

The interpretation of the results can be done by examining the reduced U and VT matrixes.

$U_{reduced}$:

CHAPTER 10 CASE STUDY ON SYMANTEC ANALYSIS

```text
U_reduced = np.array([
    [-0.577, 0.0],
    [-0.577, 0.0],
    [-0.577, 0.0],
    [0.0, -0.577],
    [0.0, -0.577],
    [0.0, -0.577],
    [0.0, 0.577],
    [0.0, 0.0],
    [0.0, 0.0],
    [0.0, 0.0],
    [0.0, 0.0],
    [0.0, 0.0],
    [0.0, 0.0],
    [0.0, 0.0]
])
```

- Each row in $U_{reduced}$ corresponds to a term from the term-document matrix.

- Each column represents a latent topic or concept.

- Terms related to smart TVs (picture quality, sound, TV): The first three rows have negative values in the first column, suggesting that the first column captures the smart TV concept.

- Terms related to earbuds (sound quality, comfortable, bass, noise-canceling): The next four rows (4-7) have either positive or negative values in the second column, indicating that the second column captures wireless headphone feature concepts.

CHAPTER 10 CASE STUDY ON SYMANTEC ANALYSIS

- Terms related to smartwatch (battery life, fitness tracking, watch, heart rate monitor, fitness tracker): The final set of rows shows very small or zero values in the reduced U matrix.

Conclusions

SVD, therefore, helped eMart automatically extract key topics from customer reviews, even when the reviews have used varied language. This automated topic discovery enables them to make data-driven decisions to improve their business.

- **Topic popularity**: By summing the singular values associated with each topic, eMart can understand which product is being discussed more frequently.

- **Customer concerns**: By analyzing the terms most associated with each topic, they can identify key customer concerns. For instance, if "battery life" is a strong term in the "smartwatch" topic, they know it's a crucial factor for customers.

- **Product improvement**: These insights can be used to enhance product design, address specific customer pain points, and refine marketing messages.

As you can be understand from this eMart example, the SVD has successfully identified two key underlying topics: smart TV features and earbuds features. eMart can now use these two topics to better understand customer preferences and improve product offerings.

CHAPTER 10 CASE STUDY ON SYMANTEC ANALYSIS

Dimensionality Reduction

Dimensionality reduction is the process in which the complexity of the semantic data is simplified. It simplifies the process of data analysis while retaining the important points of the data, making the whole process more efficient. Dimensionality reduction is achieved by lowering the rank of the matrix, which eliminates irrelevant features that can introduce noise and mitigates sparsity, thereby improving performance. It improves the efficiency of computing large datasets by speeding up data analysis. By merging dimensions associated with similar terms, the technique can recognize synonymy and latent relationships between words.

For example, while going through customer feedback, instead of looking at each word, if the focus is on the most frequently used words, the most frequently used positive words, or the most frequently used words criticizing the product, or neutral words, it helps understand the overall sentiment of the customers much better.

Cosine Similarity

After reducing matrix dimensions via SVD, LSA compares documents in a transformed semantic space using cosine similarity. This technique measures the degree to which documents align based on their vector representations. Similar documents have cosine values closer to 1, whereas dissimilar ones have values near 0. This measure is particularly valuable in fields such as data mining, natural language processing, and machine learning.

Mathematically, cosine similarity is defined as follows.

cosine similarity $(A, B) = (A.B)/|A||B|$

- $A \cdot B$ is the dot product of vectors A and B.

- $|A|$ and $|B|$ are the magnitudes (or norms) of vectors A and B.

The value of cosine similarity ranges from –1 to 1.

- 1 indicates that the vectors are identical (0 degrees apart).
- 0 indicates orthogonality (90 degrees apart), meaning no similarity.
- –1 indicates that the vectors are diametrically opposed (180 degrees apart).

To calculate cosine similarity, first, the dot product of the two vectors is to be calculated. This should be followed by the calculation of the magnitude of each vector. Then, the dot product is to be divided by the product of the magnitudes to arrive at the cosine similarity. Let's try to better understand this with the help of an example.

Consider two statements represented as vectors based on word frequency.

- **A**: "Smart TV has good picture quality"
- **B**: "Smart TV has good sound quality"

First, we convert these documents into numerical vectors (Table 10-3) based on their unique words.

Unique words: [Smart TV, has, good, picture, quality, sound]

CHAPTER 10   CASE STUDY ON SYMANTEC ANALYSIS

Table 10-3. *Convert these documents into numerical vectors*

Unique Words	A	B
Smart	1	1
TV	1	1
Has	1	1
Good	1	1
Picture	1	0
Quality	1	1
Sound	0	1

Vector of A = [1,1,1,1,1,1,0]
Vector of B = [1,1,1,1,0,1,1]

1. Calculate the dot product.

$$A.B = (1\times1)+(1\times1)+(1\times1)+(1\times1)+(1\times0)+(1\times1)+(0\times1) = 5$$

2. Calculate the magnitude of each vector.

$$|A| = \sqrt{1^2+1^2+1^2+1^2+1^2+1^2+0^2} = \sqrt{6}$$
$$|B| = \sqrt{1^2+1^2+1^2+1^2+0^2+1^2+1^2} = \sqrt{6}$$

3. Calculate the cosine similarity.

$$\frac{(A.B)}{|A||B|} = \frac{5}{\sqrt{6}\times\sqrt{6}} = 0.833$$

The cosine similarity (0.833) indicates that A and B have a high degree of similarity in terms of the words they use, even though they describe different aspects of a smart TV.

CHAPTER 10 CASE STUDY ON SYMANTEC ANALYSIS

LSA Applications

- **Concept-based search** enhances search engines by retrieving relevant results even if exact keywords differ.

- **Automated document categorization** groups text into thematic clusters.

- **Information retrieval** improves indexing by associating documents with semantically similar words.

- **Text summarization** extracts key points from lengthy texts.

- **Subjective question evaluation** assesses written responses based on their semantic content.

- **Literature-based discovery** finds hidden connections between research topics.

Among the other methods of text analytics interpretation are explicit semantic analysis, word embeddings, neural network-based models, ontology-based models, and the bag-of-words model.

10.3 Role of Semantic Analysis in Enhancing Customer Experience: A Case Study Fujitsu's Kozuchi AI Agent

10.3.1 About Fujitsu

Founded in 1935, Fujitsu Limited, a global leader in information and communication technologies (ICT), has been innovating in ICT for the last nine decades. A long list of pioneering achievements and innovative products has helped Fujitsu become a leading company in the ICT industry. From telecommunications equipment manufacturers making switching systems and transmission equipment to support telephone service networks, Fujitsu also found use for these systems in the transmission of computer data.

Two decades later, the company developed FACOM 100, Japan's first practical relay-type computer, leading to Fujitsu growing to become the country's leading computer manufacturer. Sensing opportunity, Fujitsu forayed into a wide variety of computers, including business computers for office work automation and small-sized, full-featured personal computers. Fujitsu also began developing software that included Japanese word processors specifically designed for creating documents.

In the early 1990s, Fujitsu recognized the need to address operation-specific issues faced by consumers and began providing comprehensive solution services, offering complete support from consulting to system construction and management.

In the new millennium, ICT started playing a more dominant role in all spheres of our lives, leading to the creation of much more advanced smartphones in the second half of the first decade. This ushered in an era that provided another opportunity for Fujitsu to contribute to business and society through high-performance and high-capacity data centers, servers, and mobile networks supporting smartphones. Apart from selling high-technology products, Fujitsu worked toward solving customer's problems. By bringing the Internet into every aspect of society, Fujitsu launched a new business strategy under the slogan "Everything on the Internet" in 1999.[7]

Over the last ninety years, Fujitsu has been resolving consumer problems and has made immense contributions to society through its technology solutions. With the advent of AI, the company has been working toward creating a rewarding and secure networked society, thereby bringing about a prosperous future that aims to fulfill the dreams of people globally.

CHAPTER 10   CASE STUDY ON SYMANTEC ANALYSIS

10.3.2 AI Initiatives of Fujitsu

On October 23, 2024, Fujitsu launched the Fujitsu Kozuchi AI Agent, a proprietary cloud-based AI service that can engage in numerous AI services by working independently and in collaboration with humans.

Kozuchi, in Japanese, means "magic hammer," referring to a legendary Japanese hammer that can create anything wished for. Kozuchi AI Agent represents AI capabilities, which Fujitsu claims can fulfill any business requirement. The following seven AI services are offered by the Kozuchi AI Agent (see Figure 10-1).

- **Fujitsu Kozuchi Generative AI**: Like other GenAI tools, it enhances human productivity and creativity by acting as a bridge between humans and computers for natural language and unstructured data.

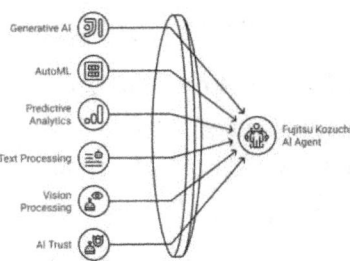

Figure 10-1. Fujitsu Kozuchi AI Agent (this pictorial representation has been created by the authors using napkin.ai)

- **Fujitsu Kozuchi AutoML**: This service helps generate high-precision machine learning models in a very short span of time. Fujitsu Kozuchi AutoML helps automatically generate AI functions for customer businesses when provided with business issues in natural language. It has the ability to convert business

455

issues into the most suitable mathematical models. The mathematical model is then converted into an appropriate AI function for the customer's business.

- **Fujitsu Kozuchi Predictive Analytics**: This service creates high-accuracy demand forecasting models using available data to cater to the changing requirements of customers for various products and aligning it automatically without the need for any human being to intervene.

- **Fujitsu Kozuchi for Text**: This service helps analyze text after it has been digitized using NLP techniques. Later in this chapter, we focus on Fujitsu Kozuchi for Text, along with other services of the Fujitsu Kozuchi AI Agent, to understand how the AI Agent performs semantic analysis that goes beyond text analytics.

- **Fujitsu Kozuchi for Vision**: This service converts visual elements through a list of 100 pre-trained basic actin models and behavior recognition rules to recognize complex human behaviors.

- **Fujitsu Kozuchi AI Trust**: This service allows verification of AI fairness, quality, and security with just a few clicks from a web browser. It also provides the much-needed service of quality and ethical standards met by the AI tools. It improves the AI literacy of the users, helping them assess accuracy, fairness, copyright compliance, data management, and misuse prevention.

- **Fujitsu Kozuchi XAI**: This service deciphers the causal relationships behind AI-generated results. Through comprehensive computation of huge datasets, Kozuchi

CHAPTER 10   CASE STUDY ON SYMANTEC ANALYSIS

Explainable AI ensures that no meaningful causal relationships are missed, thereby delivering reliable insights every time.

10.3.3 Fujitsu Kozuchi for Semantic Analysis

Fujitsu Kozuchi is a proprietary cloud-based AI platform offering wide-ranging solutions across seven AI domains to help business transformation. This section examines one critical analytical function of the platform: semantic analysis capabilities. Semantic analysis capabilities enable organizations to derive meaningful insights from unstructured textual data.

Fujitsu Kozuchi's approach to semantic analysis goes beyond basic text processing by integrating with other services in the Kozuchi platform, thereby creating an all-inclusive ecosystem that can handle large volumes of unstructured text data. This service demonstrates Fujitsu's remarkable advancement from interactive to proactive AI capabilities, excelling at transforming abstract conversational concepts into actionable tasks. It integrates the proprietary processing logic of the Kozuchi AI Agent with advanced AI models, including Takane, a large language model for enterprises that offers the highest Japanese language proficiency in the world,[8] Fujitsu AutoML, and also leverages Microsoft's Semantic Kernel. This open-source development kit enables users to easily build AI agents and integrate the latest AI models into their C#, Python, or Java codebase.

Fujitsu Kozuchi enhances business decision-making and boosts productivity. This integration helps Kozuchi to execute complex semantic interpretations of natural language inputs, beyond the literal content to the contextual meaning, purpose, and relationships between concepts expressed in text, all of which are achieved without any human input.

To understand how the Kozuchi AI Agent works, let us look at the following example, where the Kozuchi AI Agent is attending a meeting with some humans and listening to the conversations between humans. It proactively gives its suggestions.

10.3.3.1 Exhibit X.1: Kozuchi AI Agent Attending a Meeting: An Example

Miriam Graham: *"Today, I'd like to review business trends. Recently, I've been hearing that deals in Asia are not going well."*

Alex Wilber: *"No, I think Asia is fine. I've heard we lost a few small deals, but I don't think it's affecting the big picture."*

Miriam Graham: *"I know Dan, the head of Asia, has said as much, but apparently, the numbers show otherwise."*

Alex Wilber: *"Oh, you're right. The numbers are different. Maybe the large deal we recently secured hasn't been accounted for yet. Let me check with Dan right away."*

While the two executives were talking, Kozuchi AI Agent, who was unobtrusively present in the meeting, came up with additional data. It analyzed available data and proactively directed the team to perform sales analysis in the Asia region. The result presented clear, logical reasons for discrepancy across the two divergent views. The AI agent comes up with the following.

Fujitsu Kozuchi AI Agent: *"Let's get a more accurate understanding of the current situation by checking the actual number of business negotiations and changes in sales figures in the Asia region. The graphical representation showed that the sales revenue in the Asia region became half in 2024 as compared to 2023."*

The AI Agent does not stop at this. Instead, it conducts an in-depth analysis to examine what is happening in other regions. For example, it provides a graphical representation of sales revenue for the same quarter of both the current and previous financial years for all other regions in which the company operates. This way, much better-informed decisions can be made during the meeting. Depending on the discussions, Kozuchi may also assist with other data to determine the reason behind this revenue decline. Kozuchi handles all these tasks autonomously, supporting the uninterrupted flow of the meeting and contributing to increased productivity.

10.3.4 Business Applications of Fujitsu Kozuchi and Its Superiority over Other Similar Tools

Kozuchi's semantic analysis capabilities are built around Takane, which enables it to perform intelligent semantic interpretation of natural language inputs. As mentioned, the semantic capabilities extend beyond literal context to emphasize contextual relationships between entities, implied meaning, intent, and conversational nuances, besides highlighting domain-related terminologies and jargon.

On the other hand, Kozuchi's semantic analysis has its uniqueness in its autonomous operational abilities. Exhibit X.1 illustrates how this AI Agent actively analyzes ongoing discussions, identifies key points and action items, and selects appropriate data and models to provide specific inputs without even being asked to do so. This is a major difference between other reactive query-response systems and the proactive semantic insight of Kozuchi AI Agent. As shown in Exhibit X.1, the Kozuchi AI Agent can interpret meeting discussions in real-time without any human intervention, identify information gaps or differences of opinions/ interpretations by humans present in the meeting, offer insights to address those uncertainties by proactively analyzing all relevant data and data sources and offer actionable task lists based on its grasping of important semantics used during the conversation.

The Kozuchi AI Agent can also integrate "knowledge-graph extended retrieval-augmented generation (RAG)" software[9] into its text analysis process. Linking knowledge graphs to textual data helps retrieve company knowledge in a structured manner, improving logic-based inferences and contextual understanding. This is achieved by "automatically generating knowledge graphs [based] on vast amounts of data, such as laws and company regulations, company manuals, and videos."[10] All of these make Fujitsu Kozuchi AI Agent a specialist generative AI for enterprises, as opposed to the general-purpose interactive large language models deployed in the consumer space.

Among the applications of Kozuchi are customer review analysis, compliance monitoring by analyzing large sets of contracts or other legal documents, improving operational excellence, and tracking market intelligence by processing expert reviews on various social media platforms or feedback from journalists in newspapers and TV. Fujitsu expects its Kozuchi AI Agent to deliver a 25% improvement in "support desk work efficiency" and a 95% reduction in the time for planning driver allocation in the transport industry.[11]

10.4 Future Trends in Semantic Analysis

Semantic analysis is evolving at a rapid pace. This growth is driven by technological innovations, increased computational capabilities, growing data complexity, cross-disciplinary applications, and expanding application domains. We now turn our focus to examining the future trends in semantic analysis. Our focus would be on applications in

CHAPTER 10   CASE STUDY ON SYMANTEC ANALYSIS

emerging areas, keeping a close eye on technological developments. The following are some of the future trends in semantic analysis.

- **Adaptive semantic frameworks or systems** that self-learn and update fundamental structures based on real-world data.[12] These frameworks are particularly transformative in activities requiring real-time process optimization and compliance assurance.

- **Ethical semantic engineering** involves applying ethical considerations, including a bias-free environment, fairness, transparency, accountability, and respect for individual privacy, to the design, development, and implementation of semantic technologies and systems. This is done so that society can utilize machines to give meaning to data and information in a manner similar to human understanding. This is extremely important, as the increasing capabilities of semantic analysis also increase the possibility of data misuse.

  The development of technology for semantic analysis and the use of the analysis should, therefore, be done ethically since not doing that can cause irreparable damage to society. To understand it better, consider, for example, the damage that can be done in highly sensitive fields like healthcare and the criminal justice system, where very high ethical standards must be maintained alongside a precise understanding of technical terminology and relationships.[13]

- **Multimodal semantic integration** performs a combined analysis of information from multiple sensory or communicative modes, such as text,

461

numbers, images, audio, and video, to create an integrated representation of data. The primary objective of multimodal semantic integration is to develop a system that can process information from diverse sources and types to provide comprehensive, accurate, and impactful insights.[14]

Conclusion

Semantic analysis is continuously evolving along multiple dimensions concurrently. Methodological advances and technological innovations have developed the capabilities of semantic analysis beyond simple text extraction to analysis based on complex reasoning. The future of semantic analysis most likely lies in hybrid approaches combining knowledge-based methods with data-driven techniques. As this domain continues to mature, we can expect AI-based technologies to become more advanced in their analysis of linguistic nuances, more efficient in their processing capabilities, and more widely applied across diverse domains. These developments are expected to collectively transform how meaning is extracted from and how language-based information systems interact with each other.

10.5 References

[1] Brynjolfsson, E., and McAfee, A. (2011). *Race against the machine: How the digital revolution is accelerating innovation, driving productivity, and irreversibly transforming employment and the economy*. Lexington, MA: Digital Frontier Press.

CHAPTER 10 CASE STUDY ON SYMANTEC ANALYSIS

[2] Sheth, Jagdish and Kellstadt, Charles. (2020). Next frontiers of research in data driven marketing: Will techniques keep up with data tsunami? Journal of Business Research. 125. 10.1016/j.jbusres.2020.04.050

[3] Plato. (1901). Cratylus (B. Jowett, Trans.). (1999). Sue Asscher

[4] Breal, Michel. Essai de Sémantique (science des significations). Paris: Hachette, 1897.

[5] Montague, Richard (1973) The proper treatment of quantification in ordinary English, in J. Hintikka, J. Moravcsik and P. Suppes (eds.), Approaches to Natural Languages, pp. 247–270, D. Reidel Publishing Company.

[6] Floyd, R. W.: "Assigning Meanings to Programs," in Mathematical Aspects of Computer Science (Proceedings of a Symposium in Applied Mathematics), edited by J. T. Schwarz, American Mathematical Society, Providence, RI, 1967, pp. 19-32

[7] https://www.fujitsu.com/global/about/corporate/history/company-milestones/index.html#anc-06

[8] https://www.fujitsu.com/global/about/resources/news/press-releases/2024/0930-01.html

[9] *https://www.fujitsu.com/global/about/resources/news/press-releases/2024/1023-01.html*

[10] https://www.rcrwireless.com/20240604/ ai-ml/fujitsu-preps-gen-ai-framework-for-specialist-enterprise-usage

[11] https://www.rcrwireless.com/20240604/ ai-ml/fujitsu-preps-gen-ai-framework-for-specialist-enterprise-usage

[12] https://www.rcrwireless.com/20240604/ ai-ml/fujitsu-preps-gen-ai-framework-for-specialist-enterprise-usage

[13] *https://www.youtube.com/watch?v=ZH2oUIiQB1E&t=33s*

[14] Kataishi, Rodrigo, The Technological Trajectory of Semantic Analysis: A Historical-Methodological Review of NLP in Social Sciences (July 07, 2024). Available at SSRN: https://ssrn.com/abstract=5022988 or https://doi.org/10.2139/ssrn.5022988

[15] Le, Q. V. and Mikolov, T. (2014). Distributed representations of sentences and documents. In Proceedings of the 31st International Conference on Machine Learning (ICML), volume 32, pages 1188–1196

[16] Kulkarni, Shailesh & Apte, Uday & Evangelopoulos, Nicholas. (2014). The Use of Latent Semantic Analysis in Operations Management Research. Decision Sciences. 45. 10.1111/deci.12095

Index

A

Active learning, 15
Adaptive semantic frameworks/
    systems, 461
Administrative tasks, 31
Agentic AI, 362
Agents, 436
AI agents
    autonomous, 429
    stock analysis agent, 430–435
ALBERT, 329, 342
AllenNLP, 295
Anaphora resolution, 279
ANN, see Artificial neural
    network (ANN)
Antonym detection, 124, 125
Antonyms, 249, 251
Artificial neural network (ANN), 340
Attention mechanisms, 359
Automated document
    categorization, 453
Automate routine business
    processes, 20

B

Bag-of-words (BoW)
    model, 163–166
Banking and financial industry
    (BFSI), 18, 19
BERT, see Bidirectional Encoder
    Representations from
    Transformers (BERT)
BERT-Base, 329
BERT-Large, 329
Bidirectional Encoder
    Representations from
    Transformers (BERT),
    328–331, 342, 343
Bound morphemes, 150
BoW, see Bag-of-words (BoW) model
Brown corpus, 59
Business domains, 80

C

CBOW, see Continuous bag of
    words (CBOW) model
Chains, 411
Character splitting, 417
Chatbots, 136, 248
ChatGPT, 2, 23, 26
Chat models, 365–369, 372, 373
ChatPromptTemplate, 372, 373
Chat words, 202, 203
ChromaDB, 425, 437

INDEX

Chunking, 230–233, 243
Class imbalance, 14
Clustering, 7–9
Coherence modeling, 279
Cohesion analysis, 301–303
Collocation, 9, 10
Complex sentence structures, 228
Concept-based search, 453
Concordance, 10
Constituency parsing, 220, 223–225
Content analysis, 20
Context analysis, 12
Context-based synonyms, 249
Context-based WSD techniques, 251
Context compression, 383
Continuous bag of words (CBOW) model, 314, 316–318, 323, 325
Continuous refinement, 32
Contrasting relation, 293
Co-occurrence matrix, *see* Term-document matrix (TDM)
Coreference resolution, 278, 281–283
Correct word disambiguation, 234
Cosine similarity, 301, 307, 309, 425, 427–429, 450–452
CountVectorizer, 40
Cross-attention, 339
Cross-lingual NLP models, 27, 28
Customer feedback, 36, 450
Customer satisfaction, 33
Customer sentiment analysis, 35
Customer service
    benefits, 36
    challenges, 34
    critical elements, 35
    NLP implementation, 34, 35
    operations, 33

D

Data accessibility, 31
Data augmentation, 248
Data connection, 411
Data extraction
    HTML documents, 68–72
    JSON files, 72–74
    PDF files, 75–78
    word files, 64–67
Data formatting, RegEx, 97
Data proliferation, 439, 441
Data streams, 16
Decoder, 339, 340
Deep learning (DL), 5, 24, 130, 441
DeepMind Perceiver, 351
Dependency parsing, 220–222, 225–227
Dictionary-based tokenization, 154, 155, 165
Dimensionality reduction, 447, 450
Disambiguation algorithm, 13
Discourse information, 292
Discourse integration
    cohesion analysis, 301–303
    coreference resolution, 281–283
    discourse parsing, 288–293

INDEX

entity tracking, 293–295
    lexical chain, 295–298
    pronouns, 278
    RST, 283–285, 287, 288
    techniques, 278–280
    temporal relation identification, 303–306
    topic modeling, 298, 299, 301
    virtual assistant, 278
Discourse parsing, 279, 288–293
Dissimilar words, 126
DistilBERT, 329, 342, 343, 345–348, 350
Distributional hypothesis, 307
Distributional semantics, 295, 307
    See also Word embeddings
DL, *see* Deep learning (DL)
Document clustering, 230
Document integration loaders, 376
Document loaders, 376
Document summarization, 24
Document transformers, 376, 416–418

E

EDA, *see* Exploratory data analysis (EDA)
Edge computing, 361
ELECTRA, 342, 343
Electronic health records, 30
Email classifier, 6, 7
Email sentiment analysis, 212
Embeddings, 339

Emojis, 192–194
    converting words, 198, 199
Emoticons, 194–199
    converting words, 196, 197
Emotional AI, 29
Encoder, 339, 340
Entity tracking, 279, 293–295
Ethical semantic engineering, 461
Euclidean distance, 307, 308
Explainable NLP, 28, 29
Exploratory data analysis (EDA), 139
Extractive summarization, 23

F

Facial expressions, 350
Feature extraction, 13
Feedback loops, 401
Feed-forward neural network, 340
Fetch stock data, 436
Few-shot templates, 385–390
File handling, 416
File loaders, 413–416
Free morphemes, 150
Frequent words, 183, 184
Fujitsu
    AI services, 455
        AutoML, 455
        generative AI, 455
        predictive analytics, 456
        text analytics, 456
        trust, 456
        vision, 456
        XAI, 456

Fujitsu (*cont.*)
  business applications, 459, 460
  business strategy, 454
  concept, 453
  in-depth analysis, 459
  integration, 457
  meeting, AI agent, 458
  operation-specific issues, 454
  text processing, 457
Full parsing, 231, 232

G

GAN, *see* Generative adversarial networks (GAN)
GenAI, *see* Generative AI (GenAI)
Generative adversarial networks (GAN), 342
Generative AI (GenAI), 2, 356, 387, 388
  applications, 360
  LLMs, 358, 359
Generative pre-trained transformer (GPT)
  applications, 359–361
  defined, 356
  future trends, 359–361
  LLMs, 358, 359
  RNNs, 356–358
Genism, 84
Gensim, 299
Global vectors for word representation (GloVe), 325–328

GloVe, *see* Global vectors for word representation (GloVe)
Google Vision transformer (ViT), 350
GPT, *see* Generative pre-trained transformer (GPT)
GPT-3.5, 406, 408
GPT-3.5-Turbo, 374–376
GPT-4, 406, 408
GPT-based models, 328–330
Grammar correction, 213
Gutenberg corpus, 58, 60

H

Hallucinations, 23, 27, 359, 360
Healthcare, 19
Hidden Markov Models (HMMs), 340
High-quality syntactic parsing, 228
HMMs, *see* Hidden Markov Models (HMMs)
Homographs, 252
Homophones, 251, 252
HTML documents
  data extraction, 67–72
  metadata, 67
  parsing, 67
  structured and unstructured text data, 67
HTML tags, 200, 202
Hugging Face, 84, 295, 329, 331, 334
HumanMessagePromptTemplate, 373, 374
Hyponyms, 254

INDEX

I

ICT, *see* Information and communication technologies (ICT)
Inferential statistics, 439
Information and communication technologies (ICT), 453
Information extraction systems, 214
Information retrieval, 12, 442, 453
Input parsing, 390–396
Internet of Things (IoT), 360
IoT, *see* Internet of Things (IoT)

J

JSON files
    data extraction, 72–74
    properties, 72
    structured data format, 72

K

Keyword extraction, 11, 129–131, 147, 152, 420–424
Keyword searches, 20, 21
Knowledge-based methods, 235
Knowledge morphemes, 151

L

Lancaster stemmer, 168, 169, 171
LangChain
    AI agents, 429–437
    applications, 362
    business data preparation, 376, 378, 379
    chat models, 365–369
    customizing inputs and handling outputs, 400–404
    data connection
        chains, 419–424
        document loading, transformation and integration, 412–418
        file loaders, 413–416
    defined, 361
    deployment and scalability, 412
    GPT-3.5-Turbo, 374–376
    guiding AI responses, 386, 388–390
    LLMs, 362, 370, 372, 373
    model I/O, 364
    modular structure architecture, 362
    and OpenAI, 383–385
    OpenAI APIs, 362
    parsing and serialization, 390–400
    prompt and few-shot templates, 385–390
    serialization methods, 365
    setting up environment, 363, 364
    structured and unstructured data, 411
    switching, 405–408
    text embeddings, 380–383

LangChain (*cont.*)
    text embeddings and vector databases, 424–429
    tools, 411
Language generation tasks, 254
Language translation, 132, 138, 142–144
Large language models (LLMs), 2, 23, 26, 27, 328, 358–360, 362
    advantages, 405
    chains, 419–424
    chat models, 372, 373
    LangChain, 370, 372
    limitations, 405
    optimum resource management, 405
    role of, 436
Latent Dirichlet allocation (LDA), 277, 298
Latent semantic analysis (LSA)
    applications, 453
    code demo, 311–313
    concept, 442
    cosine similarity, 450–452
    dimensionality reduction, 450
    patterns, 310
    steps, 310
    SVD, 443–449
    TDM, 310
    term-document matrix, 442, 443
Latent semantic indexing (LSI), 442
Law professionals, 19
LDA, *see* Latent Dirichlet allocation (LDA)

Lemmatization, 116–119, 147, 150, 189–192
    challenges, 171
    resource-intensive process, 171
    spaCy, 174
    TextBlob, 175
    WordNet, 172, 173
Lexical analysis, 276
    challenges, 208
    data preprocessing
        chat words, 202, 203
        converting text to lowercase, 179
        lemmatization, 189–192
        machine learning, 206, 208
        removing emojis, 192–194
        removing emoticons, 194–199
        removing frequent words, 183, 184
        removing HTML tags, 200, 202
        removing rare words, 184, 186
        removing stop words, 181, 182
        removing unwanted punctuation, 180, 181
        removing URLs, 199, 200
        spelling correction, 204, 205
        stemming, 186–188
        YouTube comments spam detection, 176, 178, 179
    language components, 149

lemmatization, 171–176
morphological analysis, 150, 151
stemming, 166–171
tokenization (*see* Tokenization)
units, 149
Lexical chain, 280, 295–298
Lexical resources, 54, 55, 61, 63, 107
Lexical semantics
   antonyms, 249, 251
   antonyms detection, 259, 260
   business problems, 255, 256
   concepts, 273
   contexts, 247
   homographs, 252
   homographs detection, 261, 263
   homophones, 251, 252
   homophones detection, 260, 261
   hyponyms, 254, 265–267
   polysemy, 253
   polysemy detection, 263, 265
   synonyms, 248, 249
   synonyms detection, 258
   tokenization, 258
   tutorial, 257
Literature-based discovery, 453
Litigations, 19
Log file analysis, 99, 100
Logic-based inferences, 460
Logistic regression, 6
Long short-term memory (LSTM), 341

LSA, *see* Latent semantic analysis (LSA)
LSI, *see* Latent semantic indexing (LSI)
LSTM, *see* Long short-term memory (LSTM)

M

MABs, *see* Multi-author blogs (MABs)
MacBook model, 234
Machine learning (ML), 4, 5, 85, 130, 206, 208, 303, 441
Machine translation, 12, 22, 138, 142, 211, 214, 273, 308, 340
Masked language modeling (MLM), 343
Medical research professionals, 30
Medical text analysis
   accuracy and reliability, 32
   administrative efficiency, 31
   data accessibility, 31
   healthcare systems, 32
   languages, 30
   predictive analytics, 32
   streamlining clinical research, 30
   use cases, 33
Memory management, 411
Microblogging, 80
Minimal informational value, 182
Mission-critical events, 16
ML, *see* Machine learning (ML)

MLM, *see* Masked language modeling (MLM)
ML operations (MLOps), 87
MLOps, *see* ML operations (MLOps)
Model accuracy, 130
Model I/O controls, 364
Morphemes, 149, 150
Moses tokenizer, 163
Multi-author blogs (MABs), 80
Multimodal BERT (ViLBERT), 351
Multimodality, 335
Multimodal NLP, 350, 351
Multimodal semantic integration, 462
Multiple-lingual models (MLMs), 27, 28
Multi-query retriever, 382
Multi-word entities, 233
Multi-word expression (MWE) tokenizer, 161–163

N

Named entity recognition (NER), 11, 86, 132, 135–138, 230, 233, 234, 243, 293
Natural language processing (NLP)
  applications, 2, 16–23, 47
  business case
    customer service and interaction, 33–36
    medical text analysis, 30–33
  challenges, 3, 14–16
  cross-lingual and multiple-lingual models, 27, 28
  definition, 2
  domain, 46, 47
  emergent areas, 29, 30
  explainable, 28, 29
  fields, 2
  global data generation, 3, 4
  implementation, 54
  innovations, 24
  legal community, 46
  leveraged technologies, 33
  limitations, 3
  LLMs, 26, 27
  overview, 1
  predictive capabilities, 54
  Python, 37–45
  reasons, 3, 53, 54
  signaling system, 3
  techniques
    clustering, 7, 8
    collocation, 9, 10
    concordance, 10
    stop word removal, 11, 12
    text classification, 5, 6
    text extraction, 10, 11
    WSD, 12, 13
  text analysis, 4
  text corpora, 54–63
  training, 25
  transfer learning, 25, 26
  use cases, 17, 18
  web scraping, 78–80

INDEX

Natural Language Toolkit
(NLTK), 10, 84
    antonym and synonym
        detection, 124, 125
    interfaces, 107
    keyword extraction, 129–131
    language translation, 142–144
    lemmatization, 116–119
    model training, 108
    NER, 135–138
    POS tagging, 132–134, 217
    professionals, 108
    sentence segmentation,
        119–121
    stemming, 113–116
    stop word removal, 110–112
    text corpora, 57
    text summarization, 144–146
    tokenization, 109, 110
    WFD, 121–123
    word similarity
        calculation, 125–127
    WSD, 127–129
    WSG, 139–141
Natural language understanding
(NLU), 34
NER, *see* Named entity
recognition (NER)
Neural networks, 6, 130, 235, 314, 337, 341
NLTK, *see* Natural Language
Toolkit (NLTK)
NLU, *see* Natural language
understanding (NLU)

NMF, *see* Non-negative matrix
factorization (NMF)
Non-negative matrix factorization
(NMF), 299
Noun phrases (NP), 230–232
NP, *see* Noun phrases (NP)

O

Online messaging, 202
On-page optimization, 21
OpenAI, 383–385, 397, 399, 400
OpenAI APIs, 362
OpenAI CLIP, 350
    multimodal
        understanding, 351–353
    text-to-image search, 353, 355
Output parsing, 390–396
Over-stemming, 166
OWL, *see* Web Ontology
Language (OWL)

P

Parsing
    input, 390–396
    output, 390–396
Parsing techniques, 213
    constituency, 223–225
    defined, 220
    dependency, 220–222, 225–227
    grammatical relations, 225
    hierarchical structure, 225

INDEX

Part-of-speech (POS) tagging, 132–135, 211
   CMU tagger, 216
   example, 216, 217
   libraries, 215
   NLTK, 217
   roles, 214
   spaCy, 218, 219
   speech parts, 215
Part-of-speech tagging, 11
PDF files, data extraction, 75–78
Penn Treebank, 163
Polysemy, 235, 253
Porter2 stemmer, *see* Snowball stemmer
Porter stemmer, 166
Pragmatic analysis
   discourse integration (*see* Discourse integration)
   distributional semantics and word embeddings, 307–334
   hierarchy, 276
   NLP applications, 275, 276
   overview, 277
   social/situational context, 275
Predictive analytics, 32
Prepositional phrases (PP), 231, 232
Pre-trained models, 25, 26
PP, *see* Prepositional phrases (PP)
Prompt templates, 385–390
Punctuation-based tokenizer, 158, 159, 166

PyMuPDFLoader, 413
Python
   building vectors, 40, 41
   counts for first five entries, 42–45
   data, 38–40
   example, 41, 42
   generic NLP pipeline, 85–87
   handling string (*see* String)
   libraries, 83, 84
   NLP tasks, 84
   RegEx (*see* Regular expressions (RegEx))
   target variables, 37
   training data, 37

Q

Quantum mechanics, 7
Query-response systems, 459

R

RAG, *see* Retrieval-augmented generation (RAG)
Random forests, 6
Rare words, 184, 186
RDF, *see* Resource description framework (RDF)
Reason relation, 293
Recurrent neural networks (RNNs), 341, 356–358
RegEx, *see* Regular expressions (RegEx)

INDEX

RegexpTokenizer, 156
Regular expressions (RegEx), 11
   cleaning string, 100
   data formatting, 97
   defined, 91
   linguistic nuances, 156
   log file analysis, 99, 100
   metacharacters, 92
   methods, 92–94
   search and replace, 96
   search pattern, 92
   text normalization, 103, 104
   tokenization, 101–103
   validating phone numbers, 95
   word counting, 97, 98
Resource description framework (RDF), 441
Resource management, 405
Response times, 36
Retrieval-augmented generation (RAG), 425, 460
Reuters corpus, 59
Rhetorical structure theory (RST), 279, 283–285, 287, 288
RNNs, *see* Recurrent neural networks (RNNs)
RoBERTa, 342, 343
RST, *see* Rhetorical structure theory (RST)
Rule-based methods, 303
Rule-based techniques, 340
Rule-based tokenization, 156, 157, 165

S

Scikit-learn, 108, 299
Search engine optimization (SEO), 21, 129, 211
Search engine results pages (SERPs), 21
Self-attention mechanisms, 337–339, 357, 358
Semantic analysis, 213, 276
   chunking, 230–233
   coding tutorials, 242–247
   concepts, 229, 439
   defined, 212
   definition, 440
   formalization, 440
   Fujitsu's Kozuchi AI agent (*see* Fujitsu)
   future trends, 461–463
   lexical (*see* Lexical semantics)
   LSA (*see* Latent semantic analysis (LSA))
   NER, 230, 233, 234
   programming languages, 440
   similarity technique, 230
   standards, 441
   symbols of ideas, 440
   techniques, 212
   term-document matrix, 236, 238
   text messages, 441
   text processing pipelines, 267–272
   TF-IDF, 230, 238–242
   WSD, 229, 234, 235
   See also Syntactic analysis

INDEX

Semantic Web, 441
Sense inventory, 12
Sentence segmentation, 119–121, 147
Sentence splitting, 417
Sentence tokenization, 109
Sentiment analysis, 5, 12, 15, 34, 55, 86, 110, 132, 164, 308, 342, 349, 420–424
sent_tokenize() method, 151
SEO, *see* Search engine optimization (SEO)
Seq2Seq models, 356
Serialization, 365, 396–400
SERPs, *see* Search engine results pages (SERPs)
Shallow parsing, 230
Singular value decomposition (SVD), 442
  applying, 446
  data collection and preprocessing, 444
  data-driven decisions, 449
  dimensionality reduction, 447
  interpretation, 447, 449
  products, 444
  term-document matrix, 443, 445
  unstructured data, 443
Skip-gram models, 318, 319, 321–323, 325
Smart query helper, 383–385
Snowball stemmer, 167, 168
Social biases, 29
Social media, 212

SpaCy, 84, 215, 218, 219, 246, 295, 303
SpaCy lemmatizer, 174
Spacy tokenizer, 163
Special disambiguation algorithms, 229
Speech recognition, 251
Spelling correction, 204, 205
StanfordNLP, 215
Statistical methods, 229, 235, 250, 254
Stemming, 113–117, 147, 150, 186–188
  Lancaster, 168, 169, 171
  root forms, 166
  Snowball/Porter2, 167, 168
Stock analysis agent, 430–435
Stop words, 11, 12, 110–112, 147, 181, 182
String
  accessing characters, 89
  cleaning, 100
  creation, 88
  defined, 88
  methods, 90, 91
Subjective question evaluation, 453
Subword tokenization, 163
Summarization, 420–424
Support vector machines (SVMs), 340
SVMs, *see* Support vector machines (SVMs)
SVD, *see* Singular value decomposition (SVD)

INDEX

Synonym detection, 124, 125
Synonyms, 248, 249
Syntactic analysis
    challenges, 228, 229
    defined, 212
    fragments, 212
    NLP tasks, 212
    parsing techniques, 220–225
    POS (*see* Part-of-speech (POS) tagging)
    skills, 211
Syntax trees, 214
SystemMessagePrompt Template, 373

T

TDM, *see* Term-document matrix (TDM)
Temporal relation identification, 280, 303–306
TensorFlow, 108
TensorFlow Multimodal Toolkit (TF-MMT), 351
Term-document matrix (TDM), 230, 236–238, 244, 310, 442, 443, 445
Term frequency (TF), 130
Term frequency–inverse document frequency (TF–IDF), 86, 130, 230, 238–242, 244–247, 310
TextBlob, 84
TextBlob lemmatizer, 175

Text classification, 5, 6, 86, 110, 132, 164
Text clustering, 8
Text corpora, 107, 119
    Brown, 59
    categorisation, 54
    data processing, 55
    definition, 54
    Gutenberg, 58, 60
    lexical resources, 55
    NLTK, 57
    NLTK lexical resources, 61, 63
    Reuters, 59
    Seaborn library, 56
    web and chat text, 60
Text embeddings, 379, 424–429
    vector store and retriever, 380–383
Text extraction, 10, 11
Text normalization, 103, 104
Text processing pipelines, 267–272
Text similarity, 425, 427–429
Text summarization, 11, 23, 129, 132, 139, 144–146, 442, 453
Text-to-speech conversion, 251
TF, *see* Term frequency (TF)
TF–IDF, *see* Term frequency–inverse document frequency (TF–IDF)
TF-MMT, *see* TensorFlow Multimodal Toolkit (TF-MMT)
Time-based connection, 303
TinyBERT, 329, 342

Tokenization, 101–103, 109, 110
  BoW, 163–166
  dictionary, 154, 155
  MWE, 161–163
  punctuation, 158, 159
  rule-based, 156, 157
  Tweets, 159, 161
  whitespace, 152, 153
Token splitting, 417
Topic modeling, 280, 298, 299, 301
Traditional machine learning, 26
Traditional word embeddings
  BERT and GPT, 328–330
  fixed-size vectors, 314
  GloVe, 325–328
  Word2Vec, 314–323
Transfer learning, 25, 26, 349
Transformer models, 23, 24
Transformers
  architecture, 338
  BERT, 342, 343
  DistilBERT, 343, 345–348, 350
  elements, 339, 340
  evolution, 340, 341
  multimodal NLP, 350, 351
  OpenAI CLIP, 351–353, 355
  self-attention mechanisms, 337
  terminology, 338
Tweet tokenizer, 159, 161, 166

U

Under-stemming, 166
URLs, 199, 200

V

Vector databases, 424–429
Vector store, 379–383
Vector store retriever, 379–383
Verb phrases (VP), 230–232
Voice assistants, 248
VP, *see* Verb phrases (VP)

W

Web Ontology Language (OWL), 441
Web scraping, 56, 78–81
WFD, *see* Word frequency distribution (WFD)
Whitespace tokenization, 152, 153, 165
Wikipedia, 349
Word cloud generation (WSG), 132, 138–141, 147
Word counting, 97, 98
Word embeddings, 295, 301
  case studies and applications, 331–334
  cosine similarity, 309
  defined, 307
  Euclidean distance, 308
  examples, 307
  LSA (*see* Latent semantic analysis (LSA))
  NLP tasks, 308
  traditional, 314–330

INDEX

Word files
    data extraction, 64–67
    data integration, 63
    multiple NLP projects, 63
    processing, 63
Word frequency distribution (WFD), 121–123
WordNet, 295
WordNet lemmatizer, 172, 173
Word sense disambiguation (WSD), 12, 13, 127–129, 229, 234, 235, 243
Word similarity, 125–127
Word tokenization, 109
word_tokenize() method, 151
Word to vector (Word2Vec), 10
    CBOW, 314, 316–318, 323, 325
    defined, 314
    neural network, 314
    skip-gram models, 318, 319, 321–323, 325
Word2Vec, *see* Word to vector (Word2Vec)
Word vectors, 307
Workflow pipelines, 267
World Wide Web, 78
WSD, *see* Word sense disambiguation (WSD)
WSG, *see* Word cloud generation (WSG)

X

XGBoost, 6

Y, Z

YouTube comments spam detection, 176, 178, 179

GPSR Compliance

The European Union's (EU) General Product Safety Regulation (GPSR) is a set of rules that requires consumer products to be safe and our obligations to ensure this.

If you have any concerns about our products, you can contact us on

ProductSafety@springernature.com

In case Publisher is established outside the EU, the EU authorized representative is:

Springer Nature Customer Service Center GmbH
Europaplatz 3
69115 Heidelberg, Germany

www.ingramcontent.com/pod-product-compliance
Lightning Source LLC
LaVergne TN
LVHW010332260326

834688LV00036B/680